Intergenerational Mobilities

Drawing from work on mobilities and geographies of the lifecourse, this collection is concerned with the ways in which age, as a relational concept, is constructed and played out in mobile urban space. With studies of ageing and mobility often focusing on discrete age groups, most notably children and older people, this study seeks to fill a gap in existing literature by exploring mobility in relation to the lifecourse and generation, looking not only at the margins. Whilst some generations are increasingly mobile, others are less so and this disparity in mobility opportunity is relational as age is relational. This book addresses gaps in knowledge in relational geographies of ageing, whilst contributing to literature on mobility and transport, in particular the burgeoning field of mobility (in)justice. Here mobility is considered in its broadest sense, for example in relation to the movement or lack of movement of bodies and to computer-mediated intergenerational communications. Through focusing on urban mobile spaces, from very local spaces of medical care to global spaces of migration that are the context for intergenerational mobilities, this collection explores these interdependencies and considers ways in which intergenerational mobilities are conceptualised and researched.

Lesley Murray is Principal Lecturer in Social Science in the School of Applied Social Science, University of Brighton, UK.

Susan Robertson is Senior Lecturer in the School of Art, Design and Media at the University of Brighton, UK.

Intergenerational Mobilities

Relationality, age and lifecourse

Edited by
Lesley Murray and Susan Robertson

Routledge
Taylor & Francis Group

LONDON AND NEW YORK

First published 2017
by Routledge

2 Park Square, Milton Park, Abingdon, Oxfordshire OX14 4RN
52 Vanderbilt Avenue, New York, NY 10017

Routledge is an imprint of the Taylor & Francis Group, an informa business

First issued in paperback 2020

British Library Cataloguing in Publication Data
A catalogue record for this book is available from the British Library

Library of Congress Cataloging in Publication Data
Names: Murray, Lesley, 1966- editor. | Robertson, Susan, editor.
Title: Intergenerational mobilities : relationality, age and lifecourse / edited
by Lesley Murray and Susan Robertson.
Description: Abingdon, Oxon ; New York, NY : Routledge, 2017.
Identifiers: LCCN 2016015426 | ISBN 9781472458766 (hardback) |
ISBN 9781315589251 (e-book)
Subjects: LCSH: Spatial behavior--Social aspects. | Aging--Social aspects. |
Intergenerational relations. | Movement (Philosophy) | Urban
anthropology. | Sociology, Urban.
Classification: LCC GF95 .I67 2017 | DDC 304.2/3--dc23
LC record available at https://lccn.loc.gov/2016015426

ISBN: 978-1-4724-5876-6 (hbk)
ISBN: 978-0-367-66807-5 (pbk)

Typeset in Times New Roman
by Taylor & Francis Books

Contents

Figures

Tables

Contributors

Katia Attuyer is a qualitative social scientist, with interests in social geography and urban policies. She is a member of the Co-motion research team at the Centre for Housing Policy (CHP) at the University of York, investigating the links between mobility and well-being among older people. Prior to joining CHP, she worked at the University of St Andrews as a Teaching and Research Fellow and was previously a member of the LATTS research center (University of Paris-Est), conducting post-doctoral research on the financialization of the property sector and its implications for urban development.

Mark Bevan is a Senior Research Fellow at the Centre for Housing Policy at the University of York. Mark, who began his career at the Centre for Housing Policy, rejoined the Centre after a successful period working in housing research at the University of Newcastle, and served as Acting Director for most of 2011. Mark has maintained a keen interest in housing issues in rural areas throughout his career, including work on the private rented sector, the role of social housing, the housing and support needs of older people, the impact of second and holiday homes on rural communities, and rural homelessness. Mark has also examined a range of issues in relation to housing and later life, including the meaning of home for people in niche housing markets.

Josep Blat has been Professor of Computer Science at Universitat Pompeu Fabra since 1998, where he was founder and director of the Polytechnic School and the Department of Information and Communication Technologies. He heads a research group on interactive technologies for HCI, e-learning and 3D graphics. He got his degree in mathematics from the Universitat de València (1979), his Ph.D. from Heriot-Watt University, Edinburgh, in 1985, and was post-doc at the Université Paris–Dauphine. Josep was director of the Department of Mathematics and CS at the Universitat de les Illes Balears from 1988 to 1994.

Karen Croucher is a Research Fellow at the Centre for Housing Policy at the University of York. Karen joined the CHP in 2000 and has continued to develop one of CHP's major research themes – housing and later life, as well as drawing on her previous experience of health research to explore the links between health, housing and neighbourhood. Karen has undertaken projects on: services to support older people to live independently in the community, end of life care, older people living in the private rented sector; the housing needs and choices of older people; retirement villages, sheltered housing and extra care housing; and housing for older people in rural areas. Recently completed work includes the National Evaluation of the Handyperson Programme for the Department for Communities and Local Government (DCLG), the development of a good practice guide to support the development of lifetime neighbourhoods also for DCLG, and work on end of life care in extra care housing.

Hadrien Dubucs is a Lecturer in the Department of Geography and Planning, Sorbonne University Paris, where he is a member of the Space, Nature and Culture research group. He has been working on the project 'Contemporary Italian Migration in Europe' with Thomas Pfirsch (University of Valenciennes) and Camille Schmoll (University of Paris VII) since 2012.

Susan Möller Ferreira is a Postdoctoral Fellow at the LICEF Research Center (Laboratory of Cognitive Computing and Learning Environments) at Télé-Université of Quebec, Canada. She holds a PhD Cum Laude in Computer Science and Digital Communication by Universitat Pompeu Fabra (2015). As part of her PhD research, and collaboration with international projects, Susan conducted ethnographic research with older people with different cultural backgrounds for five years in Spain, Brazil and Denmark. Her current research interests are Human-Computer Interaction, ICT for Development, Game-based learning, older people and children.

Christian E. Fisker is the Academic Coordinator of Gerontology in the Chang School at Ryerson University, where he is part-time faculty in the School of Urban and Regional Planning. He is also Vice President of Planning and Development at Chartwell Retirement Residences, Canada's largest owner and operator of seniors' housing. He carried out his Ph.D. (2011) entitled 'End of the Road?: Loss of (auto)mobility among seniors and their altered mobilities and networks – A case study of a car-centred Canadian city and a Danish city', at Aalborg University. Fisker is a member of the Ontario Gerontology Association, the Canadian Association on Gerontology, the Ontario Professional Planners Institute and the Canadian Institute of Planners, and has co-edited a collection: *Technologies of Mobility in the Americas* (2012).

Rose Gilroy is Professor of Ageing Planning and Policy in the School of Architecture, Planning and Landscape, Newcastle University. From a practice background in social housing she has a long standing interest in older people and their environments. She is currently researching on Co-Motion and My Place, both funded by EPSRC under the Ageing and Mobility call. Recent research has included older people's everyday lives in Wuhan, China; and co-housing as an alternative housing model. She is active in

Newcastle's age friendly city action group and works closely with older people's activist groups in Newcastle.

Dave Harley is a Lecturer in psychology at the University of Brighton. His experience and education span both the technical and human spheres. An electronic engineer by trade, David's first degree was in psychology, with subsequent work in community care in the areas of mental health and learning disabilities. He recently completed his Ph.D. in the areas of human–computer interaction (HCI) and accessible design. During this period he researched older people's appropriation of computers and the Internet and aspects of intergenerational design. David has also worked in industry for some years as a user experience consultant working with organisations such as the NHS, Macmillan Cancer Support, and Random House Books. He is currently involved in his own research looking at mobile phone use in public spaces.

Laura Johnson is an international development professional with over ten years' experience working with national and international organisations in Africa. Laura has previously spent six years in Sierra Leone managing child protection and health programmes, and is now based in the UK working with ChildHope UK, managing partnerships and programmes reaching some of the world's most at risk children in East Africa and Sierra Leone.

Vicky Johnson is a Principal Research Fellow in the Education Research Centre at the University of Brighton. She has over twenty years of experience as a researcher and consultant in social and community development, both in the UK and internationally. Vicky's key focus of research and publication is in the field of children and young people's participation, with experience and expertise in broader community development, education, inclusion, and environmental issues. Vicky has led research, programmes and partnerships in Africa, Asia and Latin America for international organisations and provided expert advice for a range of UN and government departments. She has also developed programmes of community research with local authorities, the NHS, government regeneration programmes, and non-governmental organisations in the UK.

Karel Joyce Kalaw recently acquired her doctoral degree in Social Gerontology at Miami University, Oxford, Ohio, USA. Her dissertation on the return experience among overseas male Filipino workers from the Gulf areas explored the intersection of migration and age and extended the utility of Filipino indigenous methodologies. Her research interests include migration, rural ageing and grandparents-raising-children.

Thomas Klinger is a Lecturer in the Department of Human Geography at Goethe University, Frankfurt am Main, Germany, and member of the working group for mobility research. He was trained in geography, macroeconomics and political science at Trier University and at the University of Edinburgh. His research interests are the biographical and cultural dimensions of daily travel. In his recent research he examined the travel behaviour of people who encounter a new mobility culture after a residential move.

Okari Boniface Magati is an experienced development worker specialising in developing monitoring and evaluation systems and carrying out research on street connected children and health programmes. He has over six years' experience of development programmes and research designs for children and youth interventions across East Africa. In his current work with Pendekezo Letu, Kenya, he is instrumental in developing model and operating procedures, grant making and program development for their work with children living and working in the streets. Boniface holds a Masters in Epidemiology from the College of Health sciences (JKUAT), Nairobi and various professional courses aligned to M&E and research systems.

Lesley Murray is a Principal Lecturer in Social Science in the School of Applied Social Science, University of Brighton, UK. Lesley previously worked as a transport researcher/strategic planner in London government, focusing primarily on the mobility needs of marginalised groups. She is a trans-disciplinary researcher whose interests centre on the social and cultural aspects of mobilities, and has written on gendered mobilities, children's mobilities and mobile and visual methodologies and methods. She has co-edited two other collections, *Mobile Methodologies* (Palgrave Macmillan, 2010) and *Researching and Representing Mobilities: Transdisciplinary Encounters* (Palgrave Macmillan, 2014).

Thomas Pfirsch is a geographer based at the University of Valenciennes. His research interests are in in urban and social geography, especially the Italian cities of southern Europe. His work focuses specifically on mobility and spatial practices of the upper classes, as well as family dynamics in urban space. He has been working on the project 'Contemporary Italian Migration in Europe' with Hadrien Dubucs (Sorbonne University) and Camille Schmoll (University of Paris VII) since 2012.

Miriam Ricci is a Senior Research Fellow in the Department of Geography and Environmental Management at the University of the West of England (UWE), Bristol. Her main research interest is on the relationship between urban mobility and social justice, in the context of technological, cultural and social change. In particular, she has recently conducted a qualitative study of youth representations of transport disadvantage and social exclusion in a deprived neighbourhood of Bristol. Miriam is also collaborating with her colleagues at the Centre for Transport and Society, UWE, in other research projects, including impact and process evaluation of the Local Sustainable Transport Fund in the West of England and a Government-funded demonstration of autonomous vehicles in Bristol (Venturer). Prior to pursuing an academic career, Miriam worked in the private and public sector, including the European Parliament Directorate offices in Luxembourg.

Valeria Righi holds a PhD in Information and Communication Technologies by Universitat Pompeu Fabra. Her research focuses on exploring the role of technologies and participatory design to foster older people's engagement in civic contexts. As part of her participation in EU-funded projects, she co-designed and evaluated prototypes of services and technologies with over 300 older people, public bodies, cultural associations and third sector organisations. Her doctoral research introduced new approaches for tackling the ageing phenomenon in urban cities by promoting design across-age and conceptualising older people as active members of communities.

Susan Robertson studied architecture at the universities of Bath and Westminster and obtained a research master's in Cultural Geography at Royal Holloway University of London. Her practice work involves architectural projects with Denys Lasdun and in her own practice. She is Senior Lecturer in the School of Art, Design and Media at the University of Brighton, leading the M.A. in Architectural and Urban Design. Sue's research is concerned with the relationship between architecture and corporeal mobilities.

Andrea Rosales is a postdoctoral researcher in the Internet Interdisciplinary Institute at the Universitat Oberta de Catalunya. Her current research focuses on seniors and playful interactions with mobile technologies. How do seniors use mobile technologies? How does this differ from younger adults? How do older people learn how to use technologies? And, what new technologies do they imagine would be useful in their lives? She moves between ethnographic explorations, log data analysis, co-design and evaluation processes.

Sergio Sayago has been conducting research in the field of Human-Computer Interaction with older people since 2004. He has conducted five years of fieldwork, combining ethnography and participant observation, in adult educational centres and computer clubhouses in several EU countries. He was a visiting lecturer at Universitat de Lleida during the period 2014–2016, and a postdoc at Universidad Carlos III de Madrid (Spain, 2012–2014, Alliance 4 Universities Fellowship) and at University of Dundee (Scotland, 2010–2012, Beatriu de Pinós Fellowship). He holds a PhD in Computer Science and Digital Communication by Universitat Pompeu Fabra (2009).

Camille Schmoll is a Senior Lecturer in Geography at the University of Paris. Her research interests are in gender and migration, feminist geographies and qualitative research methods. Camille has been working on the project 'Contemporary Italian Migration in Europe' with Thomas Pfirsch (University of Valenciennes) and Hadrien Dubucs (Sorbonne University) since 2012.

Elaine Stratford is Professor and Director of the Peter Underwood Centre for Educational Attainment at the University of Tasmania. Elaine works in the borderlands between cultural and political geography, island studies and educational policy. She is particularly interested in thinking about ways in which, individually and collectively, we might enable more people to flourish through the life-course. Elaine has recently published more on this subject in *Geographies, Mobilities, and Rhythms over the Life-Course: Adventures in the Interval* (Routledge, 2015). Elements of the chapter in this present work are drawn from the book or indebted to it.

Rebecca Tunstall is Centre Director and Joseph Rowntree Professor of Housing Policy at the Centre for Housing Policy at York University. Rebecca has wide-ranging research interests and expertise across housing studies, social policy, and applied social research. Her principal areas of work have been social housing, neighbourhoods, and inequality, which since 2001 have been pursued as a member of LSE's ESRC Centre for the Analysis of Social Exclusion (CASE) and from 2003 as a fellow of the Brookings Institution in Washington, DC. For the past fifteen years Rebecca has worked as team leader, team member and sole researcher on projects for clients including: Communities and Local Government (CLG), the Joseph Rowntree Foundation (JRF), the Nuffield Foundation, the Housing Corporation, the Homes and Communities Agency, the Tenant Services Authority, the National Housing Federation, the Scottish Government, individual social landlords, and others.

David Walker is a Research Fellow in the Overseas Development Institute. He is a Social Development specialist, with over 10 years of experience in aspects of gender dynamics, childhood poverty and child protection (including gender-based violence) and linking evidence to policy processes within a variety of institutional contexts. David has experience in a variety of NGOs and research institutions, with in-depth fieldwork across South Africa, Ethiopia, Malawi, Uganda, Rwanda, Zambia, Papua New Guinea, Indonesia, Sri Lanka, Malaysia and Thailand, and a broader geographic focus on Sub-Saharan and Southern Africa.

1 Introduction

Conceptualising intergenerational mobilities

Lesley Murray

Douglas Coupland's novel *Generation X* is subtitled 'Tales for an accelerated culture', referring to the speed of living, of consuming and of moving of a particular generation. This is an acceleration of affluence including mobility affordances such as cars, computers and mobile phones. As Henseler (2013: 1) suggests, 'Generation X is a cohort with personal and political experiences that have marked the way they look at the world and they live in the world'. It is a generation as much determined by mobilities as the 'Y' and 'Z' generations that followed and the 'baby boomer', 'wartime' and 'pre-war' generations that preceded. The movements of people, objects, knowledge, communications and ideas, in their defining moments in history, shape generations. But at the same time these generational cultures are not solely productive of our mobility experiences and similarly they are not solely defined by their associated mobile contexts. Rather, people travel through life, back and forward in a lifecourse in which generational cultures and identities intersect with myriad other social identities in time and place. Generations imbibe, contest and reproduce the prevailing cultures of mobilities. These cultures are sometimes place-specific but often move between places. Generational identity and belonging are dependent on number of factors, and so a particular generation with a particular mobility culture is unlikely to be homogeneous according to age. Generation is a mobile social category that is lived in time and space. Hence relations between generations will be complex, multifaceted and dynamic.

This collection, which emanated from a session on intergenerational mobilities at the Royal Geographical Society with Institute of British Geographers Annual International Conference (London, August 2014) is concerned with the ways in which generation, as a relational and fluid concept, is constructed and played out in relation to other generations through mobility. It seeks to address gaps in knowledge in relational geographies of ageing, whilst contributing to literature in mobility and transport, with an emphasis on inequalities and unevenness. In this collection, mobility is considered in its broadest sense (Cresswell 2006; Sheller and Urry 2006), although here more often to the movement, or lack of movement, of bodies that are aged and ageing. Mobilities are determined through constructions of social groups according to age; for example childhood, parenthood and older age. There is emphasis also on the contexts in which these mobilities are produced, the mobile spaces, from very local spaces of medical care and urban encounters to global spaces of trans-national migration. These embodied movements and their spaces and times, together with the broader constellations of mobilities, determine inter-generational mobilities. This collection explores these interdependencies and considers ways in which intergenerational mobilities are conceptualised and researched. A mobilities approach, as I have argued elsewhere (Murray 2015a: 302) 'precipitates a transdisciplinary and intergenerational approach to the understanding of ageing and mobility' as 'an approach that is relational in respect to the co-production of age and space and the construction of age across generations'.

Age, relationality and lifecourse

Age has long been a concern of scholars interested in transport and mobility, with particular attention given to the mobilities of children and older people (for an overview see Barker et al. 2009 and Murray 2015a). However, these studies are often predicated on relational and generational understandings of age. Children's mobilities are often researched in relation to their relative independence, with a number of studies suggesting that the diminution of independence is a matter of great concern (Hillman et al. 1990; Shaw et al. 2013). However, such approaches have been critiqued for their underlying assumptions in relation to generational (adultist) constructions of childhood (Mikkelsen and Christensen 2009; Murray 2015a). Intergenerational mobilities between older people and children is often considered conflictual. For example, Wixey et al. (2005) found that 'fear of travel when buses are full of rowdy school children' was a barrier to travel for other generations. For the Mayor of London (GLA 2008: 6) 'free travel for kids has brought a culture where adults are too often terrified of the swearing, staring in-your-face-ness of the younger generation'. There are a number of correlations between children and older people's mobilities, which are often conceptualised in relation to speed. The accelerating world in which we live is geared towards generations that are not too young or too old. Children are too immobile, they are 'hypomobile' (Murray 2009), due to increasingly sedentary leisure activities. Older people are too slow in relation to speedy roads and speedy pavements. However, as Cresswell argues, 'one person's speed is another person's slowness' (2010: 21). Adopting a generational approach to understanding the overlapping and intersecting mobilities of people of different ages can help reveal this.

In a hypermobile society (Adams 1999; Phillipson 2004), age segregation excludes people from spaces that are available to others and/or disempowers according to generation. Whilst some generations are afforded increasing mobility, others are offered less. Mobilities are unequal according to age. Hopkins and Pain (2007) critique studies that focus on the margins of age through the generational categories of childhood and older age. However, most literature on age and mobilities relates to particular ages and particular 'margins of age' such as older age (Nordbakke and Schwanen 2014; Ziegler and Schwanen 2011; Musselwhite et al. 2015) because it is here that mobilities can be most marginalising and unjust. People in younger and older generations are often relegated to the edges of normative mobilities (Murray 2015). There is more to say in terms of the margins of age and how they intersect with generation, particularly in relation to imaginative and aspirational mobilities (Murray 2015; Murray and Mand 2013), embodied mobilities (Degen and Rose 2012), and relationships between memory and mobilities (Hoelscher and Alderman 2004).

Closely aligned to studies of age is the notion of lifecourse, which has attracted some recent attention in transport and mobility (Clark et al. 2014). Lifecourse and mobilities are culturally bound. The lifecourse is the path or journey of life; as Green (2010: 1) argues, it is 'about people's lifelong experiences from birth to death'. It is to be negotiated, travelled. Lifecourse approaches focus on individual histories (for example see Edmonston 2013 on migration). Bailey (2009) highlights work that uses 'mobility biographies' to illustrate the intersections of mobility and generational practices of production and reproduction such as employment, partnership and childbirth (citing Kulu 2008). So a lifecourse approach to mobilities can be used to explore the ways in which past events such as moving house or having children impact on current mobilities. Whilst this is important, we also need to look at the relational experiences of generations. Lifecourse approaches place individuals in historical context. Whilst social aspects can form part of a lifecourse approach, using for example Elder's (1994) notion of 'linked lives', they tend to underplay the complex web of interrelationships and the ways in which the construction of identities around generation and lifecourse are significant in shaping mobilities. Nevertheless, the lifecourse approach allows us to look at the in-between as well as the established. As Edmonston (2013) suggests, it is based on 'trajectories, transitions, turning points and timing'. It is often more difficult to contemplate the in-between in terms of generation, but this to an extent is related to its more ephemeral aspects. For generations are not static categories that form particular phases of personal biographies, but rather roam according to social, political and cultural contexts. Generation is fluid, relational and impactful.

The concept of lifecourse, therefore, emerges throughout the collection, as a temporal context in which intergenerational mobilities are situated. Often, studies of the lifecourse follow a linear path through various life stages from childhood, youth, middle adulthood, older age and death (for example Green 2010). This is resisted across the collection. Lifecourse is constructed as a movement in time rather than space. However, the contributors to this book are concerned with the ways in which space and time, as co-constituted and indissoluble, produce generational and lifecourse experiences. The challenge is therefore as much about disentangling these terms as it is about exploring their intersections with mobilities. The lifecourse is fluid and intersected rather than comprising immobile categories of age (Bailey's 2009). As Hockey (2009: 229) argues, there is a distinction between relatively static, established discourses of lifecourse and generation, 'the standardized life course' and more fluid and unstable 'age-based identifications'. Whilst accepting this critical approach to lifecourse and generation, in researching generational and lifecourse experiences it is often necessary to use these lived categories as, for example, in the studies of children's and older people's mobilities discussed above. A number of the chapters in this book follow this approach. They do so, however, with an awareness of their relationally, intersectionality and dynamic interdependencies, with an attentiveness to the ways in which generational and lifecourse identities are both co-produced and dependent on the interactions of age and social space.

Generational mobilities

Generations are often considered to be distinct stages or phases of the life-course. However, even with an acceptance of the non-linearity of the life-course and the uneven transpositions between particular stages and discrete generations (Green 2010), this conceptualisation and the relationality of these terms are messier than the generation-as-life-stage approach. The concept of generation sits uneasily within the lifecourse. This is because social experiences can be dependent not only on the lifecourse and its historical, cultural, spatial and political context, but also on generation and the ways in which generational identities and ascriptions are determined by these contexts. A more fluid and connected notion of the lifecourse and a relational approach to generation is needed. Hence Vanderbeck's (2007) call for more research on extrafamilial intergenerational relationships that embrace the complexity of age-related issues, and calls in both gerontology and geography/urban studies for a reconsideration of the relationality of space and age (Andrews et al. 2012; Bailey 2009; Hopkins and Pain 2007; Schwanen and Páez 2010). Vanderbeck (2007) puts forward a case for a 'less compartmentalized' approach to generational issues, which he argues remain under-researched. He (2007: 202) asks the question: 'How does space facilitate and limit extrafamilial intergenerational contact and the nature of intergenerational relationships?' A number of chapters in this collection address this, with a concern for the ways in which intergenerational relations produce and overcome mobility injustice. Riley and Riley (2000) argue that intergenerational integration can only be achieved through the breaking down of structural barriers and the bringing together of different age groups. To this end Pain (2005) worked with the Office of the Deputy Prime Minister to prepare a set of recommendations for the promotion of intergeneration practice in the UK, arguing for a bottom-up participatory approach, where different age groups are incorporated into the decision-making process.

Engaging with the concept of generation precipitates a more relational and critical scholarship. As Mannheim (1952) argued in his seminal text 'The problem of generations', generations are more than the product of biology but are associated with a 'location' in a continuum where members of a group become associated through shared experiences within a particular socio-political and historical context. Hence members of a certain generation may come to inhabit the same social space with different practices depending on their accumulated experiences or histories of mobilities. Of course generational experiences are not homogeneous but intersect with a range of social identities and diverse cultural contexts. There are also generations within generations (Hjorthol et al. 2010). Mannheim argued that particular socio-political events that are experienced by a generation produce a generational consciousness that, coupled with the internal antagonism that is produced through this diversity, has the capacity to bring about social transformation. As Vanderbeck (2007) argues, different generations have their own normative values and practices because of the particular socio-economic and political contexts within which they are born. Such thinking can be transposed to mobilities, where particular generations adopt conflictual mobile practices. Like gender and class, the characteristics of generation can be excluding in different social, spatial and temporal contexts (Alanen 2001; Pain 2001). As certain spaces are assigned to specific generations – spaces of education to children and young people and spaces of the home to older people – so too mobile spaces become marked by generation, from skate parks to vintage car rallies. It is the intersections within generations and between generation and space that produce particular conflicts and the potential for social transformation.

Intergenerational mobilities

As discussed, scholarship that cuts across disciplines often uncovers challenges and inventive ways to address them. Mobilities is a transdisciplinary approach that spans disciplines including sociology, geography, anthropology and transport studies, and offers new insights as a result. Intergenerational mobilities looks to literature in the geographies of the lifecourse (Bailey 2009; Hockey 2009; Hopkins and Pain 2007); studies in mobility biographies (for example Clark et al. 2015); lifecourse approaches to mobilities (Edmonston 2013); and in mobility histories (for example Mom 2011). 'Mobilities' is a burgeoning field of academic study, which encompasses both an acknowledgement of historical studies of the productive capacities of space and movement; and a freshness in perspective in relation to their understanding and transformative capacities (Adey 2009; Cresswell 2006; 2010; Edensor 2010; Hannam et al. 2006; Jensen 2009; 2010; Merriman 2009; Sheller and Urry 2006; Urry 2007). Whilst the associated proliferation of literature includes studies of age and mobilities, there is less attentiveness to the more relational (Murray 2015) and arguably political conceptualisation of 'generation'. It is argued that adopting an intergenerational mobilities approach allows a more comprehensive understanding of how age is produced in certain spaces; how in turn age produces these spaces; and how age is experienced in relation to other ages. For example, appreciating the ways in which older people's experience of travelling by bus in the morning can be understood in relation to children's experiences of bus travel to school. It follows that an understanding of children's experiences of the same journey illuminates older people's experiences. Mobility cultures are generational, based on the situated experiences of people of a similar age, experienced according to the construction of that particular age and the discourses around it at a particular time in history. As this transdisciplinary and international collection shows, these mobilities give rise to generational cultures across diverse scales, from local to global. Studies in lifestyle mobilities, such as Wu and Zu's (2015) research on lifestyle migration in China, illustrate the ways in which these different scales of mobility are connected through generation. They demonstrate the importance of generation in both disrupting patterns of mobilities and in determining patterns of lifestyle migration where 'family is often the reason for moving out and moving back' (ibid.: 3). A number of chapters in this collection similarly illuminate these inter-scalar interdependencies.

It is argued here that framing age according to intergenerationality and in relation to the lifecourse reveals unequal mobilities. The interactions between generations produces uneven distributions of power. We are therefore concerned with a politics of mobility, where 'mobilities are both productive of … social relations and produced by them' (Cresswell 2010: 21). Cresswell argues that mobilities produce this unevenness according to three relational aspects: material movement, representation and experienced practice. This conceptualisation is useful in illustrating the ways in which mobilities produce and are produced by social relations according to generation. First, generational mobilities are profoundly embodied experiences, recognising that the body is socially and culturally constituted and changes over time (Calasanti 2005). This applies to the bodies of all generations and to discourses of embodiment that give rise to inequalities and injustices according to generation. Spaces and places structure the ways in which bodies materially move (Schwanen et al. 2012). Second, generation is represented through mobilities. Discourses of mobility construct ideologies of generational movement. This is not only evident in representations of physical movement but in the way, for example, younger people are associated with innovations in new technologies and older people are seen to be lagging behind. Third, generation is multi-sensorially experienced according to mobile spaces. It is practised through mobilities. For example, as discussed, speed can make some embodied experience invisible as people of particular ages become characterised by speed.

This collection brings together key academics in a number of fields including geography, sociology, psychology, transport studies and ageing studies, in seeking to address a number of aspects relating to intergenerational mobilities. There are a number of intersecting themes that have been introduced here. The contributors each critically explore contemporary trends in ageing and mobilities and the intersections of generation, mobilities and the lifecourse. The chapters also examine the intersections between age and other embodied identities such as gender. Most importantly perhaps, as this is the focus of the collections, the chapters explore how aspects of the mobilities of particular generations relate to other generations through, for example, the mobility of poverty across generations. In addressing these aspects of intergenerational and lifecourse mobilities, we are aiming to fill a gap in knowledge through bridging the divides between mobilities theory, transport studies, transport policy and practice; and revealing how intergenerational mobilities can offer wider contributions to making sense of the mobility and mobile lives of diverse social groups. Such an approach has implications not only for knowledge creation but also for societal transformation. Rather than characterising generational mobilities in opposition to the desired mobilities of other generations, an intergenerational mobilities approach follows other studies (Engwicht 2005; Davis 2003; Pain 2006; Thang 2001) in highlighting the socially transformative potential of mobilities across generations.

2 Journey to the 'undiscovered country'

Growing up, growing old and moving on

Elaine Stratford

Introduction

In succumbing to a shattering existential angst, Shakespeare's Hamlet asks who would bear the burdens of a weary life were it not for the

> ... dread of something after death –
>
> The undiscover'd country, from whose bourn
>
> No traveller returns ...
>
> (Shakespeare c.1600, Act III, Scene I, 80)

This chapter is about the geographies, mobilities and rhythms of growing up, growing old, and moving on to that 'undiscovered country'. My reflections are motivated by the question: *As we move through the life-course how do we conduct or govern ourselves and each other in order to flourish?* I want to provide a (contingent) response to that question by reference to two artistic interpretations about the life-course that resonate with scholarship on mobility and the geohumanities in which some of my work is positioned. I begin by considering flourishing as a form of conduct. In *The Ethics*, Aristotle argues that everything we think, all our actions, and all our practices are intended for some higher purpose: this end is, in itself, 'the chief good' (Book I:1). Accordingly, the virtuous capacity to flourish (*eudaimonia*) moves well beyond gratuitous pleasure and warrants practical wisdom (*phronesis*).

Flourishing thus understood is both a means to, and an end that derives from, conducting oneself to leave a tangible and worthy heritage or legacy. How people address this ontological puzzle is the subject of much work in the creative arts and humanities concerned with the task 'know thyself' (Cosgrove 2011). On this understanding, I outline and then draw from insights from the geohumanities and from mobilities scholarship to undertake a close reading of two art works. One is Sandy Nicholson's (2011) *0 to 100 Project*, which illuminates some of the intergenerational dynamics with which this volume of essays is concerned. The other is the Pharmacopoeia collective's (2003) installation entitled *Cradle to Grave* (The British Museum 2014) which enables consideration of another set of dynamics: that we die at every age and stage of life and not simply at the weary end of life, as Shakespeare poignantly conveyed.

Both these multi-media works use text and image to produce powerful stories about birth, growth, senescence and death. While intrinsically interesting, here these works serve as heuristic devices: both inviting reflection upon the physical, emotional and cognitive dimensions of the body as it grows and ages; or consideration of the direction, shape, configuration, form, boundaries and structures of lived experienced in place and across space and time; or attention to scale, for example by reference to ideas of lives well-lived in proportion and degree and, indeed, justly. My aim in relation to these works is to draw out and mobilise their geographical imaginaries to reflect on questions about the life-course, flourishing and conduct.

Conceptual framework

Giele and Elder (1998, 22) define studies of the life-course as those examining 'a sequence of socially defined events and roles that the individual enacts over time'. Such studies tend to consider the trajectories of individual life experiences and life spans, the influence of different periods in time and the effects of different cohorts of age (Bynner and Wadsworth 2011; Miciukiewicz and Vigar 2013). Significantly, numerous such studies seek to elucidate and reproduce those conditions on the basis that they might foster better outcomes for individuals over their lifespan.

In *The Nicomachean Ethics* Aristotle (350 BCE-b) outlines his understanding of *eudaimonia*.[1] Throughout and in total, he argues, a flourishing life is buttressed by prosperity; forms of good fortune comprising external goods without which 'it is impossible, or not easy, to do noble acts' (Book I:8). Among these goods are friends, riches, political power, good birth, worthy children and beauty. Unjust or grasping forms of conduct are condemned (Book V:1), so too, any tendency to excess (Book X:8). Aristotle asserts that by 'virtue we mean not that of the body but that of the soul; and happiness also we can call an activity of the soul' (Book I:13). Nevertheless, *eudaimonia* is enhanced by providing for oneself because 'our nature is not self-sufficient for the purpose of contemplation [a habit he values most highly], but our body also must be healthy and must have food and other attention' (Book X:8). In *De Anima*, Aristotle (350 BCE-a) elaborates on the relationships that characterise this 'nutritive soul', the body, movement, and the nature of being. There, he observes that 'the affections of soul are enmattered formulable essences' (Book I:1); the soul is inseparable from the body and is its first reality and final cause. In both *The Ethics* and *De Anima*, Aristotle concludes that a flourishing life requires *phronesis* – a composite form of practical wisdom and judgment aiding and enhancing certain virtues (Book I:2–7). Training and habituation is key to a worthy life that is lived with deliberation, since to 'entrust to chance what is greatest and most noble would be a very defective arrangement' (Book I:9). In the context of the present work, concerned with what it means to flourish through the life-course and especially at its end, it is important to emphasise one other argument Aristotle advances in *The Ethics*. Unable to fathom the prospect that 'the fortunes of descendants and of all a man's friends should not affect his happiness at all' he declares any such eventuality 'a very unfriendly doctrine' (Book I:11). He concludes that the ancestors want to know that their ways of being in the world had good effects into the future; that they left a worthy legacy by flourishing in their own present time.

In numerous works on anatomo-politics, biopolitics, governmental regimes, and heterotopia and heterochrony, significant insights on conduct are provided by Foucault (1976; 1980; 1982; 1986; 1991). In turn, Rose (1998, 178) asks of governing, conduct, and power: 'Who speaks, according to what criteria of truth, from what places, in what relations, acting in what ways, supported by what habits, routines, authorized in what ways, in what spaces and places, and under what forms of persuasion?' Dean (1999, 17–18) observes that 'we govern ... others and ourselves according to various truths about our existence and nature as human beings ... the ways in which we govern and conduct ourselves gives rise to different ways of producing truth'. More recently, Wright (2010; 2013) has argued that deficits in flourishing are socially caused and resolvable. His agenda for change involves identifying moral principles useful for judging the worth of social institutions, and applying those principles to diagnose and critique present institutions to transform social life and realise hope-filled alternatives. For Wright, such tasks are like a voyage in which one first *maps* what is wrong in the world, then determines what motivates a need to depart it, imagines what new world is sought, and finally works out, practically, how to *mobilise* that change. Notwithstanding, he remains cognizant of the tensions inherent in the idea that utopia refers both to the good place and to no place at all.

Here, Wright's metaphors of the map and journey are fitting given my focus on flourishing over the life-course and home terrain in human geography. His ideas also dovetail with observations advanced by Cloke (2002) about how geographers should honour the connections between the ethical and political actions embedded in public intellectual endeavours and our life experiences as people, which resonates with the ideas Aristotle wrangled with. In short, Cloke calls for both 'geographically sensitive ethics, and an ethically sensitive geography' (ibid., 591) that might furnish the conditions in which to flourish.

Cloke's concern with the value of the discipline maps onto a larger question about the value of the humanities in general and the geohumanities in particular. Of the first, McDonald (2011, 290) notes that the humanities are a 'crucible where we test and judge values ... or perhaps more accurately "meta"-values', including those which are unconscious or tacit. Intrinsically, the arts have capacity to transform how we understand and live in the world. According to Martin and Jacobus (1983, 453), 'all artists, consciously or unconsciously, attempt to organize a medium into a form-content, into a structure that reveals something significant that has never been revealed'. But more, artists clarify, reveal and shift individual, social and cultural values and, in so doing, many are motivated by normative impulses. Hawkins (2011) is aware of this dynamic when she reflects upon why geographers should study art works, and points to a certain kind of politics: a connective aesthetics which, among other things, engages place and culture. So too, Luria (2012) calls for significantly more collaboration between literary critics and geographers, not least because of the potential for transformative thinking.

In one early edited collection on the geohumanities Dear (2011, 16) argues that there is a need both for artists as visionaries and 'scholar critics' to mobilise the translation of ideas to realities. In another such collection Cresswell (2011) notes the utility of work in the social sciences producing profound shifts in understanding about the meaning and reach of mobilities (on which, see Sheller and Urry 2006). Nevertheless, Cresswell argues that more effort needs to be directed to understanding 'what mobility has been made to mean and how these meanings have been authorized ... an approach informed by the humanities ... [which] has an insistently critical edge' (2011, 78–79). Here, Cresswell points to the crucial importance of the geographical imagination – geosophy – on the basis that this impulse 'feeds into all kinds of ways of making the world up ... [informing] arguments about how the world is and how it should be and how we might get from here to there' (ibid., 82).

On two grounds, the idea of getting from here to there is useful in thinking about the life-course. First, there continues to be value in unsettling prevailing metaphors of life as an inexorable trajectory from youth to agedness and enfeeblement and then on to death when, in fact, it involves all sorts of detours that should be of direct interest to geographers, not least because they reveal much about the ontologies of 'wandering off score' (after Edensor 2010). As Bergmann and Sager (2008, 8) have noted, there is a need to 'describe and explore contemporary and narrow notions of mobility [and] ... investigate and nurture their transition to richer conceptual models and practices'. Second, the geohumanities provide ways to engage in being, doing, representing, performing and critically reflecting upon these journeys and their constitutive connections with the geographical imagination as it gains expression in 'maps, photographs, paintings, films, novels, poems, performances, monuments, buildings, traveller's tales and geography texts' (Daniels et al. 2011, xxvi). I now turn to give close attention to two such geographical imaginings as they relate to understandings of the life-course.

The *0 to 100* Project

0 to 100 is the culmination of ten-day photo sessions with 300 people in Sydney and Toronto by Canadian photographer Sandy Nicholson and further engagement with 101 of those people. Their images and narratives about the life-course feature in several outputs listed on the project's website (www.sandy nicholson.com/projects/0-100). The project *places* individuals according to age *and* opens up *other spaces and times*; ways of looking at the life-course that invite reflections that defy normative temporal or spatial frames and point to Foucauldian ideas of heterotopia, heterochrony and conduct. In this way, *0 to 100* invites consideration of the geographies, mobilities and rhythms of the life-course because it explores participant responses to being *placed* at an age or stage of the life-course or their efforts to *mobilise* varied resources for varied ends.

My initial exposure to the *0 to 100 Project* app took the form of a rapid-fire nine-second slide show of head shots in front of a uniform background emphasising the unique characteristics of each participant as well as the common humanity of all. The show starts with the image of a howling infant named Robin. It ends with the visage of a smiling centenarian named Marguerite. Less than a year old, Robin says nothing but his crumpled face suggests that he carries the world's cares; in contrast, Marguerite's face is peaceful. Between the two are images of fifty-five females and forty-four males between one and ninety-nine years of age, most accompanied by snippets of conversations about the life-course. I have no access to the full conversations shared with Nicholson, and nor can I catalogue every utterance made publicly available. Instead, I have referred to those extracts he selected and which pertain to age, ageing, generational change and ideas about flourishing and conduct. Four clusters of ideas emerge.

The *first cluster of ideas* centres on how participants feel at different ages. Reid (9) relates how the 'best bit is being young and able to get lots of candy': an aspiration unlikely to sit well with Aristotle's ideas! Jacob (15) wearily observes that 'when you're a teenager, there's a lot … of prejudices against you'. Why this may be the case is captured by Aristotle (350 BCE-b, Book IV:9) when he writes that shame is not becoming to every age, but only to youth. We think young people should be prone to the feeling of shame because they live by feeling and therefore commit many errors but are restrained by shame and we praise young people who are prone to this feeling. But no one would praise an older person for being prone to the sense of disgrace since we think s/he should not do anything that need cause this sense.

In contrast, Dom (20) reports that he 'feels good … feeling a little bit more like an adult' and he equates ageing with the getting of wisdom: 'one year older, a little bit … wiser'. Sofia (23) is less sanguine, observing: 'I need to grow up quite soon. Which is a bit scary.' Nicholas tells Nicholson 'I wish I could stay (25) forever', which perhaps evinces a capacity to overcome developmental angst or, conversely, hints at a realisation that he knows there may be more such worries to come. In turn, Michael (29) feels that 'childhood's got a little bit more stretched out than my parents' generation'. Dianne reveals that she 'love [s] being 33 … I was dreading it at first, but I can be myself now … I enjoy like and I do what I want … I'm loving growing older and can't wait until I'm in my 90s … it's good, ageing.' Robert (35) reveals that he has 'a young son, and that's the best part – you know, watching him grow up'. Likewise, Jeff (40) suggests that 'the best things about being 40 are the things that come with it – career, house, wife, family, friends, all that kind of good stuff'. Lori

(47) quips '[a]s I get older; ...older ... the word "old" keeps getting pushed out, so to me, old is over 70 now'.

A *second cluster of ideas* relates to participants' aspirations, ambitions and emerging and growing sense of responsibility as they age. The youngest person to share her thoughts with Nicholson is Jaylah (3), who tells him 'I'll be a princess. A rock star. I'm going to be a princess rock-star.' Ena tells him 'I'm ten and it's very fun being ten ... I want to be a good dancer.' At 12, Lochlan reveals that it 'feels like you're goin' after girls and stuff ... and you explore new options ... you have to do all this stuff ... so much homework when you start high school... it's horrible ... I think I want to be a lawyer or write books or something'. Drawing on the idea of a 'majority', Josh notes that it is 'beautiful to be 18 ... it's the best thing you can be ... when life pretty much begins ... taking responsibility but not too much'.

Josh's observations prompted me to turn to Golombek's (2006) work on four notions of citizenship: *jus solis* – signifying individuals born in a particular country; *jus sanguinis* – marking the children of parents born in that country; naturalisation – the acquisition of citizenship by choice, where offered; and reaching one's majority, which is what Josh seems to be alluding to. All four imply rights and responsibilities that require particular sorts of conduct, but none actually assumes that children and young people are active citizens. Rather, employment, sports, education, school governance or voluntarism are engagements that are seen as preparations for adult citizenship and, according to Golombek, undermine 'the active status of children in constructing and determining their social lives, the lives of others and their surroundings' (ibid., 14). Such views uphold the idea, captured by Jacob above, that young people are 'mere' human becomings (Valentine 1996).

The pressure to 'grow up' and conduct oneself in particular ways gains expression in Dong's comments to Nicholson. At 27, he refers to 'all the time I wasted when I was in school and all these, you know, things I should have done when I had the chance', as though such options are closed to him. More confidently, Michael (30) feels that people of his age have 'a little bit of time to adventure, to explore yourself, make mistakes' but he too observes that 'time is running out to finalise the path and foundation for your life'. Michael also draws on the idea that past conduct is a powerful resource, suggesting that there is merit in using 'the experiences of the twenties [and] let certain things go'. In the final analysis, Michael thinks of his age as akin to ... 'being at half-time – are you up or down in score or out of the game ... you don't really know'. A decade and more on from this point, Savita (42) reports self-assuredly that, for her, 'the best bit is that you have certain experiences, certain wisdom, certain knowledge, and still the energy left to enjoy'.

A *third cluster of ideas* – how one's prospects may change or be threatened – emerges in comments made by Charmaine when she reveals to Nicholson that 'this is a transitional time ... I'm currently in the process of being treated for breast cancer, and that's the worst part of being 45, but the prognosis is fabulous and I'm going to be fine'. These sorts of situations perhaps urge Maria (49) to offer the following prescription: 'Look after yourself. Eat healthy. Care about yourself. Keep yourself active. Keep positive. Smile.' In fact, global and national models of present and predicted burdens of inactivity and ill health, and burgeoning services for health care suggest that ageing will be a profound fiscal challenge (United Nations, Department of Economic and Social Affairs, Population Division 2002; 2013). Notwithstanding, middle-aged and older adults in more affluent communities will enjoy substantial consumer choice in relation to health and well-being services. Biggs (1997, 555) goes so far as to submit that such people will use their purchasing powers to recode the body, a form of conduct emphasising 'the fluidity of identity choice'. Nevertheless, Biggs points out that these masks or masquerades, as he refers to them, are not equivalent to other technologies of the self that people use to enhance fitness and well-being, and literally their effects may be skin-deep. Even so, that which is 'merely' cosmetic may be profoundly influential: William (51) protesting that, at 'a certain point in my life, others will decide that my career is over, simply because of the amount of grey hair I have'. David (53) observes that the 'body just stops responding quite the same way. Sort of like a car that's just come out of warranty'. At 56, and apparently yearning for a different and younger self, Syd tells Nicholson 'what I know now with a 20-year-old body – I would be really dangerous. I would be a lethal weapon'. Ray (58) confesses 'I still feel like a teenager ... although some days I get up and certain joints ache ... but I never say to myself "maybe you're getting old, Ray, no, you're getting better, you're just wearing out, that's all"'.

Given that the body is constituted as a physical and moral project (Andrews et al. 2012; Rudman 2006), it is not surprising that over the life-course body image is seen as critically important (Baker and Gringart 2009). Time seems to engrave specific expectations upon one's conduct (Nikander 2009). For instance, it is commonplace for women's and men's magazines and social media sites to feature articles about beauty regimes, diet and exercise using decadal 'slices' from the twenties to the fifties. In relation to the sixties and later decades, silence illuminates an oppressive and double-edged process. On the one hand, strong social and cultural messages compel people to engage in quests to be 'not old' and these may be allied to other messages about health and a sense of flourishing (see Flourish Over 50, 2014). On the other hand, there is a widespread tendency to ageism (Calasanti 2005).

Nicholson's participants are not immune to this idea of inevitable decline and loss. While upbeat, Marianna (60) says 'I can still walk, run and enjoy life' and yet she is clear that she wants 'to be gone before I get old, very old ... when I cannot move: I don't like that'. Being 'gone' of course is an experience felt not by the departed but by those who remain and, poignantly, Darlene (76) confides in Nicholson that she 'didn't expect to be a widow, and I didn't really know how I would be living, but I didn't think I'd be living alone'. In this vein, Maddrell (2013, 503) reminds us that death 'is an everyday reality', arguing that there is 'considerable scope for cultural geographers to offer more insight to this universal and often life-shifting manifestation of absence'.

Assuredly, several participants understand these dynamics of decline and loss. Marlene (71) describes how 'you look in your face, and you can't believe it's you'. John (75) takes a larger environmental scan, and is glad that he is 'going to be spared global warming and runaway population growth and food shortages and all these terrible things that are coming upon us'. Another John says of 80 that 'it doesn't feel a hell of a lot different than when I was 40, I still do the same things, it just takes me longer'. Glyn (97) observes that all he plans 'to leave behind, I hope, is a good impression'. And of course, it is not just those who age who experience its effects, Ruby (6) admitting that 'when someone is old, they are really tired. Poppy is old. He can't remember things. It's okay for me. I don't know how it is for him'.

A *fourth cluster of ideas* about age, ageing, generational change, and flourishing and conduct centre on showing gratitude for one's life and its lessons, and these resonate with the idea of what is required is not only absolute virtue but also a 'complete life, since many changes occur in life, and all manner of chances, and the most prosperous may fall into great misfortunes in old age ... and one who has experienced such chances and has ended wretchedly no one calls happy' (Aristotle 350 BCE-b, Book I:9).

Hence, Errol (50) exclaims 'Oh Lord! ... I lost a co-worker at 47 ... Every day's a blessing.' Jude (65) resolves to make every year 'really count. I did before, but now I'm really going to make these ones count.' Tetiana (70) reflects that 'I'm ... more mellow ... a little bit wiser in the sense that ... I know now I can't change the whole world ...there has to be a balance'. Deborah (72) states simply that 'Old is knowing what matters and rejecting the rest.'

Aileene (79) is quoted at some length. She tells Nicholson her age is a great achievement and reveals that she has enjoyed every age for the special things each brings; she is confident that she has many years ahead of her, because she comes from a 'family of longevity and that seems to be a good thing'. Aileene also acknowledges she has much to be glad of in terms of income and travel and much to live for, including grandchildren ranging from 20 to 6. So she says 'I guess I haven't got anything to complain about with my life.' The oldest participant, Marguerite (100), is also quoted at length. She tells Nicholson 'sometimes I can't believe I'm a 100. I don't feel it, not in my head. But sometimes in my body I know I'm a 100.' Marguerite is sanguine, revealing that she worries about nothing, doing what she likes one day at a time without plans. And she says 'I've learned to be tolerant; to understand that everyone has something good in them and you look for that; and I've learned that people are generally good if you look for it'.

Brief though each of these narratives is, *en masse* they enlighten: here are many and disparate life experiences: aspirations, dreams, projections, concerns, joys, ennui, despondency, resignation and hope. Here is evidence of people seeking for themselves the requirements of full and accountable lives. Here is acknowledgment that others also govern experience: parents, bosses, databases, societal expectations, stereotypes, circumstances, systems and processes. Here is understanding that there is a need, ultimately, to embrace certain forms of conduct that witness the coming of (practical) wisdom.

Cradle to Grave

Perhaps before, but certainly by 2003, textile artist Susie Freeman, video artist David Critchley and general practitioner Liz Lee were working as a collective called Pharmacopoeia. One of the group's major works is *Cradle to Grave*, which forms a permanent installation in the centre of the Wellcome Trust Gallery at the British Museum. There, in 2011, I first encountered the 15-metre glass cabinet that houses a lifetime's supply of prescribed drugs sewn into two lengths of textile, each 0.7m wide. The perimeter of each half-length displays 'stories of men's and women's lives in captioned personal photographs, documents, and objects' (The British Museum 2014, no page) (Figure 2.1).

On each occasion I have since visited the installation, visitors surround it, showing animated interest in the items on display and the overarching narratives they convey. It would be surprising if *Cradle to Grave* elicited anything other than this fascination, for laid before witnesses is a composite storyline exhibiting profound particularities about individual lives; more generalisable insights about specific cohorts of people in developed nations; and themes that might be construed as universal which are focused upon what it means to be born, live, die and leave a legacy.

Figure 2.1 Cradle to Grave (2003). Art installation, 15m × 2m.
Wellcome Trust Gallery, the British Museum. Courtesy of Pharmacopoeia (Susie Freeman, David Critchley, and Dr Liz Lee). Reproduced with permission.

To take in the richness of the exhibit, walking around the long cabinet several times takes on the feel of a journey, an act of witnessing others' life-courses. The identities of people whose lives are laid out here are less immediately apparent than in *0 to 100*. Yet no fewer insights are possible in relation to the geographies, mobilities and rhythms of the life-course, and how they are implicated in the conduct of our lives. If one begins the aforementioned journey at that corner of the cabinet marking the start of life of the distaff or female line, assembled under the thick glass are various artefacts: a birth certificate with tiny inky-pink footprints and a photograph of a young man and a neonate gazing at one another on a hospital gurney. Across the cabinet is young Charlie being given his nebuliser treatment, aged 1, and some way beyond that a snapshot of Matthew learning to rollerblade.

Down the two cabinet sides thus unfold stories of those exceptional entities, our ordinary lives, revealed in discourses both medicalised and humanised and focused on health, well-being, connection, restraint and excess: the exhibiting of a half-filled wine glass and of cigarette butts in an ashtray testimony to this last tendency. And further down the cabinet: a medical certificate referring to Robert's back pain; an empty transfusion bag, tea-brown liquid still visible at its margins; eye glasses, dentures and hearing aids. And then, Alistair and

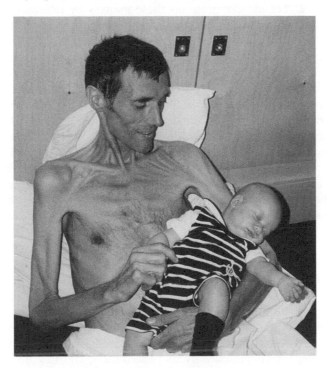

Figure 2.2 Alistair and Ann-Sophie. *Cradle to Grave* art installation (2003), 15m × 2m. Wellcome Trust Gallery, the British Museum. Work by Pharmacopoeia (Susie Freeman, David Critchley, and Dr Liz Lee) Reproduced with permission.

Ann-Sophie: a man dying in a hospital bed and the latest addition to his extended family (Figure 2.2). Beyond that, a death certificate for a plumber from Bristol: stroke, high blood pressure and cardiovascular disease, his daughter as witness and comfort in his final hours.

What is one to make of these evocations? Assuredly, our very *being* involves senescence and death, and across varied scales our lifeliness and its termination allude to questions of how we conduct ourselves in response to great pain, grief or loss. That remark stands, I think, even where relief and quiet resignation are mingled with the experience of dying, but either way at some point our bodies appear to 'betray' us and vigour moves into apparent stillness, furtively, gradually or with such speed that there is little or no chance to adjust. Sometimes this change from life to dying unto death is embedded in deeply unjust systems, processes and structures; such as arise in conditions of unremitting poverty or violence. Sometimes the change seems grievously unfair, especially when it feels that someone's life and capacity to flourish have been cut short. Sometimes, markedly among the very old, this shift appears natural and that might give it the appearance of being 'just'.

But let there be no doubt: ageing may involve deep abjection and, without the salve of intimate care from others, its terminal stages feel anything but natural or just. Let me elaborate on these matters by reference to the very young and the very old. Consider in this light a story in the populist magazine the *Australian Women's Weekly*. There, Caroline Overington (2012) reported on the experiences of a couple named Paul Murray and Siân Horstead. Their son Leo died 34 hours after his birth in Sydney in August that year because he had a vealmentous umbilical insertion in which the blood vessels in the cord ended 15 centimetres short of the bulk of the placenta. In Leo's case the condition, neither preventable nor detectable, resulted in one of the vessels breaking, leaving him compromised *in utero* and unable to recover after birth.

Overington's account partly focuses upon Deb De Wilde, a consultant obstetric social worker who attended the couple and whose job it is to 'encircle ... parents, to keep them safe and try to slow things down for them, to allow them time to be with their baby, to begin creating memories ... and to be guided in their actions not by fear, because we can be afraid of death, but by love' (ibid., 63). Overington describes how, at interview, Paul and Siân recollected Leo's simple, wonderful and utterly normal actions in the few hours in which he lived. Thereafter, with family and friends they were able to create a special service for Leo described as neither funeral nor wake but celebration of life. Of such support, in *The Ethics* Aristotle (350 BCE-b, Book IX:11) notes that 'the very presence of friends is pleasant both in good fortune and also in bad, since grief is lightened when friends sorrow with us. Hence one might ask whether they share as it were our burden, or – without that happening – their presence by its pleasantness, and the thought of their grieving with us, make our pain less'.

At the other end of the spectrum, if one lives long enough to enter the ranks of the oldest-old, one will still submit to something terminal. In the west, ischemic heart and cerebrovascular diseases prevail as chief causes of death but among those 85 years and older the third most common cause is organic (including symptomatic) mental disorders such as Alzheimer's disease. Influenza, pneumonia, renal failure and other degenerative diseases of the nervous system also emerge in the top ten causes of death among the oldest-old (United Nations, Department of Economic and Social Affairs, Population Division 2013).

Abrahamsson and Simpson (2011) provide a number of insights about what constitutes the boundaries, capacities, and thresholds of bodies that are helpful for understanding how the corporeal is subject to morbidity and mortality. They note, for example, that while limited in Euclidean space bodies can still 'extend beyond themselves ... [be] slowed down or held still' (ibid., 331). Thus, the body is limited, declines and decays, eventualities raising the predicament of when, whether and how to foster life: matters not just governmental but biopolitical, entangled in questions of conduct and the exercise of power (Dean 2002). For example, how is one to reflect upon the biopolitical status of an elderly person, without living relatives, in a vegetative state in a nursing home from whom will never come instructions a 'properly whole' patient could provide? According to Agamben (1998, 94), there exists in such contexts something akin to 'death in motion ...[bodies] kept alive by life-support systems ... [that] have the legal status of corpses'. The point is that death is a political decision raising the spectre of bare life and prompting questions about what counts as life, as worth living, and about what can be left to die. Such matters profoundly affect what it means to flourish.

In turn, for the very old, whether at home or in care, the composite process of living, becoming frail and dying represents a significant challenge. Among other things this process implicates the physical properties and internal and external arrangements of space: the functional facility that is a nursing home room; the ways in which those arrangements may give effect to meaningful places; the geographical range of inhabitants, visitors and caregivers; the patterns of travel made by each and the justifications they offer for those journeys; their use or refusal of technologies such as online goods and services; or staying in touch with family and friends.

Crucially, for my consideration of flourishing, Lord et al. (2011) describe this last transition into great age as one *that need not reduce* ageing to the status of problem. That idea seems important in *Cradle to Grave*, in which there are several instances of humorous engagement with ageing, or photographs of elderly people engaged in quiet contemplation (Figure 2.3). In particular, Lord and his colleagues refer to a process that in French is termed *déprise*, whereby daily life is rationalised to provide for shifting circumstances without any suggestion of a 'downward spiral'. In my reading, this idea of *déprise* infers a calming quietude, a bearing that underpins the capacity to flourish, and which is to be distinguished from acquiescence, a surrender that resembles capitulation.

Figure 2.3 Gertrude contemplates the future, 1989. *Cradle to Grave* art installation
(2003), 15m × 2m.
Wellcome Trust Gallery, the British Museum. Work by Pharmacopoeia (Susie Freeman,
David Critchley, and Dr Liz Lee) Reproduced with permission.

The idea of *déprise* also challenges that of frailty, defined by Lynn and
Adamson (2003, 5) as 'a fatal chronic condition in which all of the body's systems have little reserve and small upsets cause cascading health problems' (Lynn
and Adamson 2003, 5). Yet, in ways that refract back to Aristotle's ideas about
the ancestors, Nicholson and Hockley (2011) argue that frailty is *necessary* if one
is to mourn the loss of all things, including oneself, and then be able to 'invest …
in people and things that will outlive you'. Where it is afforded, then, our

scripting of the last chapter of life constitutes a space in which to write of a life well lived and, as Cowley (2013, 3) points out, the alternative is 'the source of the peculiar vulnerability to humiliation suffered by the elderly'.

Cowley's work focuses upon those who know it is time to prepare for death and seek to flourish by referring to the whole of their lives, much as Aileene was able to do in *0 to 100*. This focus pinpoints a paradox: in such circumstances flourishing is retroactive but its fulfilment is pre-emptive and deliberate, a form of conduct enacted over decades. It affords a legacy rejecting what Cowley calls the thin liberalism and physiological deficit informing prevailing ideas of ageing. In the same way, Leavy (2011, 706–708) insists upon the possibility that 'old age might have something good to offer in itself … unique, new, fresh, … its benefits exist not in spite of physical and mental limitations, but joined with them'. This revelation invites us to reinstate the archaic and absented meaning of age that is to be experienced, nourishing and strengthened (Oxford University 1971).

Conclusion

I began this reflection by asking: As we move through the life-course how do we conduct or govern ourselves and each other in order to flourish? Acknowledging that there will be diverse ways in which to address that query, here I have proposed that rich insights are to be gained by reference to the geohumanities and mobilities scholarship. I have also sought to demonstrate how creative works, in this instance the *0 to 100 Project* and *Cradle to Grave*, serve as powerful heuristic devices enabling reflection on experiences of growing up, growing old and moving on.

Underpinning these reflections has been an understanding of flourishing indebted to Aristotle, who viewed this state-of-being as symptomatic of purposeful and considered lives rendered so by the application of practical wisdom. All the same, there is a need to acknowledge that varied forms of anatomo-, bio- and governmental forms of power shape our lives in ways that cannot be anticipated, even as we pursue hope-filled alternatives to what may be dystopian conditions. The mere fact of ageing is one so complicated by these diverse forms of power and our capacity to face our maturation, senescence and demise is perhaps never fully in our grasp. Notwithstanding, it has been my aim to demonstrate that the impulse to craft purposive and fulfilling lives can be powerfully imagined, performed, written and represented, and thus reflected upon. I have attempted to do that by reading into and out from the *0 to 100 Project* and *Cradle to Grave* a range of insights on such matters. It has been my intention also to show how these works engage with journeys through the life-course in ways that emphasise the need to produce ethically sensitive geographies and manifest geographically sensitive ethics. Such labours could support efforts in the humanities wherein are considered a range of meta-values, not least those related to the life-course and the journey to the undiscovered country.

Note

1 Aristotle has been charged with dismissing the capacity of the young, slaves and women, to engage in the sort of intellectually and morally virtuous political life that he seeks to constitute as eudaimonic. Yet, as Mann (2012, 201) has suggested, Aristotle's elitism becomes less clear-cut and spaces for emancipation become conceivable if one accounts for and finds potential correctives in his 'association of the love mothers feel for their children ... with the work of laboring to produce another human being ... [an insight that invites us to consider] his fluid, constructivist conception of nature, which Aristotle clearly believes is contingent on one's function, and one's function is contingent on the activity that one takes up, or is allowed to take up'.

3 Moving between generations?

The role of familial inter-generational relations in older people's mobility

*Rose Gilroy, Katia Attuyer, Mark Bevan,
Karen Croucher and Rebecca Tunstall*

Introduction

Drawing on new data from a study of mobilities and well-being experienced through transition management, this chapter takes a relational approach to consider the value of familial inter-generational activities in the context of mobilities. The chapter draws on existing literature and on new empirical data. This includes interviews with a sample of fifty-one people aged 55+ in three sites in northern England, carried out in spring and summer 2014. The chapter concludes by considering the implications of our findings in light of the growing number of people who will age without children, and therefore whose inter-generational connections will need to be built beyond traditional family networks.

If there is a marked 'mobilities turn' in social sciences, mobilities has also played a part in creating a new paradigm of ageing. For if social science 'treats as normal stability, meaning and place' (Sheller and Urry 2006, 208) then the 'good' old age has been characterised as static, emplaced (often in the home setting) and feeding off deep roots in small places. New paradigms of ageing have emerged in which the 'third age' is built on a diversity of post-work lifestyle options and freedoms built from consumption that blur ideas about a fixed life-course with a predictable post-retirement pattern (see Gilleard and Higgs 2000). Within this, the freedom to be mobile is critically associated with independence and choice-making, with a normative expectation of auto-mobility[1] and aero-mobility for a group that grew up with the emergence of the package holiday. Within social policy, the mobility of older people is also understood as connectivity to social, intellectual and cultural stimuli, while mobility as exercise is seen as part of the recipe to keep older people active and, significant in this argument, from bearing down too heavily on pressed NHS resources.

Various commentators have deepened our understanding of the notion of mobility, and developed nuanced insights into how the concept is experienced in later life (Nordbakke and Schwanen 2014; Parkhurst et al. 2012; Prohaska et al. 2011; Ziegler and Schwanen 2011). Recent work has reviewed what is known about the different dimensions and facets of mobility, drawing upon a wide range of perspectives to inform how the term can be defined and conceptualised (Stjernborg et al. 2014). Some definitions of mobility focus on the physical ability to move limbs and to move the body, for example to walk over short distances and to be able to use public transport. 'Mobility dependency' has been defined as needing help or being unable to walk 400 yards, climb up or down stairs, or get on a bus (Ayis et al. 2006). Others measure mobility in terms of calories consumed (Sawatzky et al. 2007), or the number of trips or distance travelled. Here, mobility can operate at national or global scale, assisted by planes and other technologies, while at the smallest scale 'micro-mobilities' include 'taking a couple of steps on one's property, supervising maintenance work, conversing with neighbours' (Lord et al. 2011, 58). A number of mobility measures assess the actual 'life space' experienced such as rooms or places visited (e.g. Peel et al. 2005; Zeitler et al. 2012). Others measure resources for mobility such as access to cars and to social networks (e.g. Gagliardi et al. 2007). Alternatively, mobility can be seen as a latent capacity, and potentially as one incorporating a subjective element: for example, as 'the ability to choose where, when and which activities to take part in outside the home in everyday life' (Nordbakke 2013, 166).

Whichever definitions are used, mobility tends to vary with age. For example, only 4 per cent of those aged 16–49 have a 'mobility difficulty', but this rises to 10 per cent of those aged 50–59, 18 per cent of those aged 60–69, and 38 per cent of those aged over 70. Annual distance travelled per year peaks in the 40s and then declines. The number of trips people make per year peaks in the 50s and then declines (Department for Transport 2011). Nevertheless, Hjorthol (2013) concluded that whilst the desired level of activity may diminish in later life, the importance of mobility for older people does not diminish. However, whilst mobility is clearly affected by health (Fristedt et al. 2014), research has also emphasised a wide range of other factors including gender, household type, employment circumstances, income and access to transport services. Webber et al. (2010) encapsulated these dimensions in a model that puts forward seven life-space locations, ranging from the room in which an individual sleeps to the wider world, each of which is composed of mobility determinants related to financial, psychosocial, environmental, physical and cognitive factors. Finally, they suggest that gender, biographical and cultural influences exert an influence on all five determinants. Linked with these latter influences, Manderscheid (2014) has argued that too often movement has been conceptualised as the product of individual decisions by autonomous agents. She contests the autonomy of the solo traveller and calls for a relational approach to mobility that pays attention to how a person's choice of travel route and travel mode is shaped by the mobility practices and representations of others, belonging to their family or social network.

Thus, while mobility is often seen as determined by a combination of individual physical functions, car ownership and transport systems, in this chapter we consider the mobilities of older people through a relational approach that explores the nature and extent to which the families, generations and networks in which older people are embedded promote or inhibit their mobility (see Hopkins and Pain 2007 for a discussion of the benefits of thinking relationally about age). Mobility can be caused by, enabled by or otherwise linked to social relations, duties and motivation. In practice, social relations provide a major motive for mobility, through companionship on journeys for pleasure and by providing destinations. Social relations can also affect people's ability to act on motives for travel, for example, by providing confidence and support, as well as practical assistance such as lifts in cars, pushing a wheelchair or help with fares. The nature or absence of social relations can also play an inhibiting role, where others' wants, needs or restricted mobility may have a negative impact on mobility, or if people have no-one to visit or to travel with. Further, the attitudes of others may also either inhibit or encourage the potential for mobility.

Inter-generational relations are distinct from relations with contemporaries and peer group. The term inter-generational encapsulates a broad sweep of social relations that may be familial, but can also be non-familial in nature (including friends or other forms of social contact). In this chapter we focus on 'familial inter-generational' where the relationship can be with parents, children and grandchildren.

Study data and methods

This chapter is based on survey data and qualitative interviews carried out in 2014 with a sample of people aged 55 or over in the cities of York and Leeds and the market town of Hexham, all located in northern England. The data gathering forms part of the 'Co-Motion' project led by the University of York, with other university partners. This project is, in turn, one of seven projects supported by the Engineering and Physical Sciences Research Council's 'Design for wellbeing: ageing and mobility in the built environment' programme. The project commenced in 2013 and will complete in 2016.

Participation involved an initial questionnaire, which was followed by a qualitative face-to-face interview and then four short phone interviews spaced at 3-month intervals that ask people to comment on their everyday mobilities and sense of well-being. The final data-gathering stages will repeat the questionnaire and the qualitative face-to-face interaction. In the later phases of research, the project will be working to translate early findings into a series of practical projects, from iPhone apps to negotiation processes and crowd-sourced information, to aid mobility and well-being amongst older people. This chapter reports on both the qualitative and quantitative data gathered and analysed by early 2015 from fifty-one participants. Participants are identified with a short descriptor of the transitions experienced in 2013/2014.

The core of Co-Motion is tracking older people over time, exploring their mobility and well-being as they move through a range of critical life transitions. The participants in the study include people aged 55 and over who experienced one or more of the following key life 'transitions' in the year prior to their recruitment to the project:

- Stopping work;
- Starting/stopping being a carer (for an adult);
- Starting/stopping significant childcare responsibility (at least one day a week);
- Starting to use a mobility scooter or other mobility aid;
- Stopping driving;
- Significant loss of sight or hearing;
- Starting to live on their own (whether through divorce, separation or bereavement); and
- Moving house.

In focusing on transitions, we agree with Grenier's assertion (2012) that physical and social transitions are more significant than chronological age. The transitions were selected because of their potential to impact on mobility and well-being, as well as on social relations, and several are implicitly inter-woven with familial inter-generational relations. For example, taking on significant care of grandchildren; moving house to facilitate the giving or receiving of care and support; becoming or stopping being a carer of an adult may be inter-generational in the case of parental care.

The onset of significant sight or hearing loss may cause a change of mobility mode that may be supported by social relations; retirement may shift both its frequency and purpose, creating new possibilities for social interaction with family and friends; while starting to live alone through death, divorce or entry of a loved one into care may free or empty out time and create mobility change by loss of escort, or mode of travel where the loss is of the household's driver. Where the transition is giving up driving, the literature has explored the role of family and friends in this critical decision (Johnson [1995; 1998] explores supportive actions while Kostyniuk and Shope [1998] explore the impact of family pressure to stop driving). In considering the impact of transitions we acknowledge that mobility is also a resource to which we have unequal access (Skeggs 2004) and which may need to be re-evaluated, renegotiated and re-affirmed.

The significance of inter-generational mobility

In the pen and paper questionnaire, we asked participants to identify particular facets of the journeys that they were making. These facets included recording the most regular or frequent journeys that they made. Participants also indicated which were the longest journeys they had made 'yesterday', 'last week' and 'last year' as well as the journeys that they felt were the most important. Participants also recorded the destination and purpose of journeys. These quantitative results were followed up in the qualitative interviews to explore people's views and experiences.

Despite the relatively high prevalence of disability, limiting long-term conditions, poor self-rated health and the use of mobility aids, almost all of the fifty-one participants were physically mobile to some degree. All but one mentioned inter-generational (both familial and non-familial) social relations, and these relations clearly affected mobility by generating additional mobility, different schedules or modes of mobility – or in some cases by restricting mobility.

Inter-generational mobilities were made up of journeys with, or to meet, other generations for mutual socialising, to provide and/or receive care, journeys with other generations providing and/or receiving support on the journey itself, journeys to destinations of shared interest, both for necessities and for pleasure. Jensen et al. (2015) highlighted how travel choices can be made to maximise time together rather than be based on considerations of time and cost minimisation. For a minority they generated the most frequent journeys, the longest journeys, and the most important mobility.

Frequent journeys

For some of the participants a feature of the familial inter-generational mobility was its frequency, which accounted for the most frequent journeys for a small minority of participants, 7 out of 51. Four regular patterns involved visiting daughters or in one case 'being taken out' by a daughter, one visiting a son, and two visiting and escorting fathers. For example, Diana usually spent Sunday with her daughter or son and she usually had lunch with them at their homes or out. She usually also saw her daughter and granddaughter mid-week and they usually went shopping, went for a short walk in the park or to eat a pizza. Childcare was often on a weekly schedule (Diana/using aid/hearing loss/childcare). In another instance, the familial inter-generational mobility involved trips to see a parent. For this participant, her father lived a ten-minute walk from her home and she visited him at least three times each week. She commented, 'Taking dad out can be a pleasure; on the other hand, it depends what sort of mood he's in because he's got Alzheimer's so some days he's fine. Other days, the lights are on and nobody's in; it can be difficult' (Rebecca/became carer).

One participant, who had started caring one day a week for a grandson who lived in York, also mentioned 'going down to Devon [about 300 miles away], that's the most important, to see family' every six weeks (Grace/started childcare). However, for others many of these familial inter-generational trips, particularly those that involved long distances, were irregular.

Longest journeys

The mean 'furthest distance travelled last year' by participants was 1,270 miles and the majority of trips were international plane journeys for holidays. In most case where information was available, these were trips with partners and in a minority of cases with friends, fellow churchgoers or solo. However, in a minority of cases, these longest trips were familial inter-generational travelling with children and or grandchildren, or for another small group these were trips to visit children and/or grandchildren living in another region of the UK or abroad.

Important journeys

While only 7 of the 51 had identified familial inter-generational mobility as their most regular journey, twelve identified it as their most important journey. For most participants, these included trips to see children and grandchildren. Melissa said the most important trips had been journeys to hospital when she was having cancer treatment but after the treatment was successful, the most important trip was the weekly journey to visit and to collect her grandchild at the weekend, which she looked forward to all week (Melissa/childcare). In another example, one participant said that she and her husband 'felt a bit lost' when their son left home, and their dog died. Becoming grandparents in the past year had given them 'a new lease of life', including extra mobility at weekends when they collect the grandchild, have her to stay, take her to the park and try to be out of doors with her. They might walk to a park or take the car and the child's buggy. Having a grandchild was getting them out more, and they enjoyed taking her to new places (Diana/using aid/hearing loss/childcare). Similarly, another said 'we [he and his wife] get out and about doing things with [our granddaughter]' (Anthony/stopped work). Although Melissa and her husband made numerous other trips in the week to work, for functional reasons and intra-generational social reasons, these trips to and with the grandchild were the most important to her. One participant did not have grandchildren but carried out voluntary work to support families with young children, requiring weekly journeys (Rebecca/became carer). One said going to her allotment was the most important trip, as it was shared with her daughter and family who she met there, and provided a site of memory of and refuge from a series of bereavements (Abigail/started childcare).

One York participant said that the most important trips were 'the family ones' to their children in York, in Alnwick (about 100 miles away), and London (about 200 miles away), although the latter appeared to take place only a few times a year (Oliver/stopped work). Another also said the most important trips were keeping in touch with the family: 'we like to go and see them at least two or three times a year' (Andrew/stopped work). Beatrice referred to trips to family in Liverpool and Leicester (Beatrice/sight loss/living on own/stopped being a carer). Another travelled within Leeds and to Huddersfield about 20 miles away (Ruth/ hearing loss). Another said that her most important trip was to the post office, to send items to children living abroad (Megan/stopped driving/stopped work).

In many cases, familial inter-generational social mobility occurred which did not account for the most regular, longest or important journeys. Many of these trips involved irregular, relatively local and sometimes partly utilitarian meetings with adult children. For example, one respondent lived in a village outside York, and had a son with wife and granddaughter living in the city that they saw irregularly (Oliver/stopped work). These familial inter-generational meetings were attached to other reasons for mobility to the city, generated by social activity: 'We see them every so often. No regular thing. But for example if I go to the [social club] meeting ... he lives nearby so we'll drop in'. Socialising with and/or caring for grandchildren created diverse forms of mobility, in some cases regular if not the most regular. Becoming a grandparent generated a substantial amount and range of mobility – visiting socially, babysitting (Sean/living on own/started childcare), caring for the child in their own home with associated delivering (Daisy/started aids/started childcare) or doing the school run (Beth/moved house/started childcare). Eight participants had started or stopped significant childcare in the past 12 months.

Supporting others and receiving support

Some participants reported that they were 'looking after or help or support others due to health, disability or old age'. Most of these were looking after spouses, but a handful of these were involved in inter-generational caring for parents or other older adults, which involved mobility. Many of these made regular journeys as part of volunteering commitments, which included inter-generational social relations, either with older people, children or a wide range of ages. A small number had extensive informal caring responsibilities.

Many participants discussed getting lifts. In a minority of cases participants' mobility was dependent on or enabled by younger family members, particularly children and grandchildren, who supported mobility by taking respondents out, enabling journeys or minimising the risk or difficulty of journeys.

Many of the social relations and care relations described above are likely to have an element of reciprocity in them. One respondent was using her skills as an artist to teach a younger neighbour about painting. She sometimes accepted lifts from the neighbour in return. A small group of participants were involved in enabling the mobility of other family members, including parents and other older family members, or disabled children. For one respondent, who could not walk much beyond 20 yards, being able to drive reaffirmed his role as a 'dad', enabling him to give lifts to his young adult children (Jack/stopped work/using aid).

Social relations inhibiting mobility

The literature has emphasised the ways in which non-familial inter-generational contact can include negative encounters with different age groups, or the way in which negative perceptions of younger generations can influence mobility (Jones et al. 2013). Pain (2005, 15) underlines that, overall, older people might not be as a group more prone than other age groups to stay in because of their fear of experiencing crime. Furthermore, however, for some subgroups of older people, such as those who are frail or live in deprived areas, the presence of groups of young people in public space is strongly associated with negative feelings regarding their neighbourhood and their ability to meet their needs, including their mobility needs. These perspectives were reinforced in our study, and a number of participants discussed how the drinking culture of younger age groups shaped the timing and geography of movement, especially in relation to travel within city or town centres.

However, what was also clear from a number of participants was the way in which familial inter-generational relations – both close and fleeting – also imposed boundaries and restrictions on mobility, as well as caring responsibilities for partners. One participant said, 'I was hampered when [my daughter's] daughter was new … I looked after her … Three days and a night a week … I was spending more time in [daughter's home town, about 30 miles from home] than I was here' (Emily/started using aids/started childcare/lost hearing). In another instance a participant also highlighted not only how her travel routines were determined by the childcare needs of her daughter, but also the way in which this participant's physical impairments required careful negotiation of parts of the journey when she got to her daughter's home (Violet/using a mobility scooter). In contrast, another participant described how the need to be close to their mother-in-law, who was living with a chronic condition, shaped their mobility (Janet/using an aid). This latter case illustrates the growing trend towards familial inter-generational relations *within* later life. These narratives also resonate with the observations by Wiles (2003) with regard to the ways in which the mobility of carers can be shaped and possibly constricted. Wiles (2003) also highlighted how a concern for risk minimisation by the person being cared for could limit and constrain mobility.

In our study, this concern for risk minimisation was sometimes a feature of familial inter-generational relations, not by the participant themselves but by wider family members. The attitudes of young family members to the safety of mobility were an influence on a large minority of respondents. One participant said, 'I'm confined to barracks in the winter … I get banned from going out with my family, you know, "You don't go out in this, you stay where you are." They're frightened of me falling and slipping, doing anything. "We don't want any broken hips"!' (Ruth/hearing loss). This latter case perhaps illustrates the observation by Katz and Lowenstein (2010) on the ambivalent nature of familial social relations between generations and the tensions generated between autonomy and dependence. Such tensions are exacerbated as parents become frail and reliant on their adult children for whom they were once providers. This chimes with Gawande's (2014) observation that for ourselves we want autonomy but for those we love we demand security, which may restrict autonomy to unacceptable levels.

In line with the literature, driving in older age was or had been an issue between some participants and their younger relatives, so Victoria (given up work/stopped driving) recalled with anger her son telling her that she 'was a hopeless driver' while others had made their decision in conjunction with their children (Benjamin/using an aid) or to help their children financially:

I was driving for a living, you know, I was all over the countryside – driving was not a problem. No, I just thought, it's an expensive hobby. If you're not working. It is an expensive hobby and I wasn't going far with it just to Portpatrick and Moffatt. And so when the daughter was trading hers in and I thought, well, I'll take mine along and trade that in. That's how she got the book price.

(Walter/stopped driving)

Inter-generational familial relations affected mobility through discussions of new technology. One respondent (Rachel/stopped being a carer) was toying with the idea of getting a mobility scooter 'because it could be fun', but was being discouraged by family members who viewed scooters as being 'for disabled people'. In contrast another respondent's son had given her a mobility scooter, but this was untouched in the garage: she wouldn't entertain the idea of using it (Lily). Another had been advised by her daughter and confessed thus: 'I said I'd never have a shopping trolley, but it was one of the best things I ever bought' (Lily/stopped being a carer).

Conclusions

Our findings support the view of Pooley et al. (2005) who argue for greater attention to the impact of life-course events, including the impact of familial inter-generational networks and relations, on daily mobility. Our study has highlighted that for older people, familial social relations make an important contribution to the extent and shape of their mobility. Familial inter-generational social relations provide an important motivator for mobility and also enable mobility, through providing escorts or transport assistance. Nevertheless, the extent and nature of familial inter-generational mobility varied substantially between individuals. This depended on the size of families, the location of participants' children and grandchildren and other familial inter-generational contacts, and participants' mobility, and was to some extent circular, with social relations and mobility reinforcing each other.

However, familial social relations can also inhibit or alter mobility, through the demands of caring roles, or through direct discouragement intended to reduce the risks of travel. Thus familial social relationships can also create an important determinant of, and constraint on, mobility. Evidence from this study, in which older participants report on relations with those of other generations, supports the relational nature of mobility for adults of other ages too (see for example Jensen et al. 2015).

The discussion of inter-generational relations in this chapter has focused almost wholly on familial links. As longevity increases the spacing between the oldest and youngest societal members and increased mobility restricts opportunities for inter-generational contact, it is the 'bonds and routines of family life' (Lloyd 2008, 29) that can provide a focus for enabling generations to come together. In contrast, Uhlenberg and de Jong Gierveld (2004, cited in

Vanderbeck 2007) revealed that among their Dutch respondents only 15 per cent of those aged 80+ had weekly contact with someone who was younger than 65 and not a family member. However, it is predicted that by 2030 there will be more than 1 million people in the 65–74 age group who will have no children (McNeil and Hunter 2014; Ivanova and Dykstra 2015; AWOC 2015). If inter-generational contact is beneficial not simply to older people but in facilitating the transmission of values, skills and human capital for the benefit of all generations, is it possible to develop spaces and opportunities for positive *extra-familial* inter-generational linkages (Lloyd 2008)?

A number of specific examples highlight how these societal changes are being potentially addressed. Responding to these trends has long been a policy objective in Germany, with the spread of over 500 multi-generational houses in which previously separated services such as youth groups, day care for older people, advice centres and childcare services come together with cafes and bistros so that people of all ages can give and receive support according to their abilities and needs (McNeil and Hunter 2014). At the grass roots level a further example, an Australian project called 'Men's Sheds', has been widely embraced as providing activity and a sense of purpose for older men while giving, in many cases, younger people the opportunity to gain workplace skills and confidence (Duncan 2015). Beyond these conscious efforts to build inter-generational and peer-to-peer networks it has been argued that with the emergence of the third ager there will be evidence of positive associations for older adults through their positioning in more diverse networks. Antonucci et al. (2006, 207) suggest that 'more varied relationships including, but not limited to family, and extended to embrace significant relationships with friends and other acquaintances, will increasingly enrich the lives of these healthier, better-educated, more cognitively and physically able adults'. The challenge comes of course that this positive view of third age life leaves many behind. In challenging isolation and promoting inter-generational and community cohesion, the Campaign to End Loneliness has called for national government to reconfigure services, and challenges individuals to realise their responsibility to cherish connections and keep contributing to their communities (Age UK 2011). Part of this policy challenge comes back to enabling and sustaining how people get out and about, but our study emphasises the relational nature of mobility, with a focus on how mobility is mediated within familial structures.

The focus on common life transitions in this study usefully highlights the complexity and change that people aged 55 and over may experience in their lives. In many cases the impact of a transition on mobility was clear; however, in other cases, these transitions merely added to the complexity of people's lives, which included ongoing and changing inter-generational mobilities. Furthermore, social relations per se, and familial inter-generational relations specifically, are primary drivers that generate reasons for being mobile. But it is also the less tangible facets of people's relations with their families and others that influence and shape mobility. People's perceptions of others, as well as the attitudes that family members impart, can also inhibit or enable mobility in later life.

Note

1 According to Department of Transport statistics for 2012, 63 per cent of men and 21 per cent of women aged 85 and older held a full driving licence.

4 Breaking intergenerational transmissions of poverty

Perspectives of street-connected girls in Nairobi

Vicky Johnson, Laura Johnson, Okari Boniface Magati and David Walker

Introduction

Street-connected girls face risks both on and during their journeys to the street. They not only face discrimination in the mobile spaces they occupy in public space, but have often moved to escape from abuse and neglect in dysfunctional families. Therefore, the success of interventions to improve their well-being is dependent on understanding their vulnerabilities and the risks they face in their mobilities in the street itself, as well as from street to school. Key to this understanding is how they negotiate the spaces they occupy, including the intergenerational and intragenerational power dynamics that permeate their complex everyday lives.

The aims of the research were: to generate new knowledge on how to improve the lives of street-connected girls in Nairobi; and to share best practice on social protection interventions including rehabilitation and reintegration into education. This chapter will address two aspects of the research project 'Mitaani hadi shuleni' meaning 'from the street into school', led by the University of Brighton with international partners. First, the subjective well-being indicators of girls and their mothers (following the work of Moncrieffe 2009 and Sumner et al. 2009), and second, their journeys to the streets and the situation they find themselves in when they get there, including understanding their experiences of mobilities and the spaces they occupy (following, for example, ideas of Porter and Mawdsley 2008 on mobility and Mannion 2010 on participatory space and intergenerational performance). This research has helped to understand the complexity of the girls' lives on the streets and how interventions could contribute to changing their situation. This fits with Hanson and Nieuwenhuys' (2013) reconceptualisation of child rights that takes into account children's complex realities and social justice. The gap between the expression of rights in international agreements to their variable fulfilment at national and local levels was also therefore recognised during the research.

Whilst treating marginalised street-connected girls as active agents of change (for example Ennew 1994), their vulnerability has been further understood in order to inform interventions to improve their lives (as suggested by Mizen and

Ofosu-kusi 2013). Some of the girls who participated in this research are orphans, but others live and work with their street families. The design of the research and analysis therefore utilised concepts of intergenerational cultural transmissions (Mead 1970) and intergenerational transmissions of poverty (Moncrieffe 2009). Relationships between the girls and their parents or guardians, and the harm or abuse they experience, helped to determine what social protection interventions had made a positive difference to their lives. Furthermore, the study also demonstrated how small NGOs can be seen as catalysts for shifts in thinking and practice (Uvin et al. 2000), particularly in linking child protection and social protection systems' strengthening approaches (see Devereux and Sabates-Wheeler 2011; and Save the Children 2011).

The child-centred research was funded by the UN Girls Education Initiative (UNGEI) and led by the Education Research Centre at the University of Brighton in partnership with a national Kenyan non-governmental organisation, Pendekezo Letu, ChildHope UK and the Overseas Development Institute. Results have informed ongoing interventions made by both Pendekezo Letu, and broader policy implications and sustainability were discussed at a reference group of representatives from national ministries, international donors and non-governmental organisations.

Research context

Pendekezo Letu is a local non-government organisation that supports street-connected girls who live and work in the informal settlements in and around Nairobi. Each year, Pendekezo Letu enrols 100 street-connected girls into a ten-month rehabilitation programme, and provides psychosocial support, remedial education and life skills training. Through a holistic approach, Pendekezo Letu also provides siblings and primary caregivers with psychosocial counselling and advice about educational and livelihood opportunities. Programme participants living with HIV and AIDS are assisted in accessing medical care and encouraged to join HIV support groups. To increase sustainability, Pendekezo Letu also support the establishment and development of local child protection services. This includes setting up Community-Based Child Protection Committees to facilitate the identification and referral of child abuse cases within their area, while providing training and resources to the local ministry department responsible for children's services, teachers and juvenile justice staff.

In recent years, there has been a rapid increase in the number of street children in Nairobi, but it is difficult to understand the extent of the problem due to the lack of data to quantify the exact number. The most recent, reliable report from the United Nations (2007) estimated that there are 60,000 children living and working on the streets of Nairobi, with estimates of more than 250,000 across the country (Arthur 2013). The post-election violence that rocked the country between 2007 and 2008, and the HIV/AIDS epidemic that continues to ravage the region has seen increasingly large numbers of people, including many unaccompanied children, ending up on the streets.

There are various efforts to mitigate the street children crisis, including the City Council of Nairobi investing in a rehabilitation centre in Ruai (Machakos County). Pendekezo Letu also works with local government representatives and other non-government organisations to share information and develop new ways to provide support to street-connected girls that includes using this research to inform and influence ongoing dialogue nationally and internationally on policy and practice.

The child-focused approach to the research with street-connected girls

The research starts from the perspectives of young marginalised girls in Nairobi (see Johnson et al. 2015), placing children and young people at the centre, taking into account their identity, inclusion and interest in participating, whilst understanding their cultural and political contexts (Johnson et al. 2013). It is also rights-based (Beazley and Ennew 2006) in that it engages with children using a mix of methods, including creative ethnographic and participatory visual methods, detailed case study interviews with girls, their families and a range of local stakeholders, including governmental and non-governmental services. It also incorporates a quantitative aspect through the use of a coding system. This evidence focuses on individual case study girls and, whilst maintaining confidentiality, allows analysis through a variety of categories designed to show their differences, for example age, ethnic group, religion, family situation, and level of education (following Johnson and Nurick 2003). The methodology builds on six steps that help to create an environment to enable the researcher to safely engage children in understanding their complex everyday lives. These steps include building trust, developing ethical protocols (including anonymity) and procedures, and using creative methods (Johnson et al. 2014).

Creative methods included visualisations based on trees, rivers and roads to represent the girls' stories. For example, trees included: roots that signified reasons or root causes to explain why girls and their families became connected to the street; the trunk representing the types of support that girls found helped them to survive; and leaves providing ideas of what the girls felt could help them to improve their situation. Rivers and roads of life showed challenges and facilitators that the girls faced on their journeys to living and working on the streets. Mapping and photo narratives explained where girls and their families live and work, while also depicting areas the girls liked and did not like, or found safe or unsafe. Other methods, such as support network diagrams, ranking lines and matrices, were used to address the aspects of the research that focused on evaluating different social and child protection services and interventions. Themed analysis was assisted by the coding system, so exemplary stories could be selected to depict the complexity of the girls' lives, and lessons shared on social protection interventions including increasing educational opportunities and outcomes for the street-connected girls.

The research addressed a key question: What social protection and child protection policies and strategies can help street-connected girls in Nairobi break the intergenerational transmission of poverty and improve their well-being? Emerging from analysis within the team of Kenyan and international researchers, detailed questions explored: the relationships between rehabilitation spaces for girls and their families with longer term improvement in child protection and education; and how intergenerational and peer relationships could both engender a sense of safety and belonging, or be seen as causal factors for girls becoming more connected to the streets or working on the streets, and dropping out of school.

The research was carried out in three phases during which 213 marginalised girls were consulted in small group discussions and individual interviews, using creative methods, and forty-eight were followed up as detailed case studies. In the first phase (July–October 2014), girls' and mothers' subjective indicators of well-being were identified and previous participatory impact assessment data analysed to understand the impact of the operational programmes of the Nairobi-based organisation, Pendekezo Letu, on street-connected girls and their families. This included original participatory action research with ninety girls (aged 4–19 years) at a rehabilitation centre, and thirty-five girls and boys (aged 10–14 years) from school clubs. Group discussions were also held with mothers who were engaged in interventions to live positively with HIV and to explore income-earning opportunities to sustain their families.

Phases 2 and 3 of the research (November 2014–June 2015) were carried out to further understand the lives of street-connected girls and their families in five informal settlement areas around Nairobi. During these phases, 123 marginalised girls (aged 10–19 years) were consulted in focus group discussions. From these, the team identified forty-eight street-connected girls to engage in detailed case studies. These included the girls, their families (including parents/ guardians and siblings) and peers in the community. The girls who participated in the detailed case studies were selected in order that about one third were supported by PKL and still at school; one third had been supported by PKL but had dropped out of the programme and out of school; and a final third were unsupported and at the stage where support would be identified for them. At least one parent/guardian and one peer or sibling was interviewed for each of the girls selected as case studies. In addition, twenty local stakeholders were interviewed, including local police officers, head teachers, elders and chiefs, probation officers, voluntary children's officers, community child protection structure members and aid workers.

A team of fifteen Kenyan researchers was trained in qualitative child-centred research in Phase 2, in which they co-constructed the detailed research questions and decided which creative methods to use. During the field data gathering stage the team had regular debriefing and re-planning sessions. The team included ten university students and graduates, and five social workers from Pendekezo Letu to support them to follow up on sensitive issues raised by the girls during the research. The project was coordinated by researchers from the UK, a lead from the University of Brighton (V. Johnson), a coordinator from the international development charity ChildHope (L. Johnson), and a Kenyan coordinator based in Nairobi (O.B. Magati). There was also an adviser from the Overseas Development Institute (D. Walker). This mix of UK and Kenyan researchers provided international comparative input, as well as ensuring research capacity was built and left in partner organisations and in Nairobi.

A Reference Group of governmental and non-governmental policy makers and service providers from across Nairobi was formed to represent ministries and organisations relevant to meeting child rights in terms of the three Ps of the UN Convention on the Rights of the Child: Protection, Provision and Participation. This reference group met in Nairobi twice, once at the beginning of Phase 2 to provide input as stakeholders in the research and once at the end of Phase 3 to facilitate the opportunity for enhanced research uptake, including a consideration of how local and national policies and strategies might be influenced by using the findings.

Understanding girls' and mothers' subjective indicators of well-being

The first phase of the research provided a background to the perspectives of street-connected girls and mothers living together on the streets before more detailed information was gathered from girls about their journeys to the street and the conditions they experience on the street (see next section). The problems girls and their mothers face living on the streets include: financial poverty leading to families separating; poor health including HIV; poor access to services including education (for children), health and contraception (for young people and adults); adults not able to make good decisions; girls keeping bad company and negative peer pressure; poor self-esteem leading to risky behaviour and in some cases conflict with the law. Girls' dreams for the future included: being happy without problems; playing with friends; having a home and living with their parents; washing themselves every day; finishing or continuing with their education; getting good jobs; helping their families and other children; their parents changing their behaviour towards the girls; changing their own behaviour and becoming a role model; and having a positive influence on their peers. Ultimately, these highly mobile children imagined a less transient lifestyle and one with intergenerational dependencies.

All of the girls engaged in the research could be considered as marginalised, as they live in informal settlements and experience financial poverty. In order to understand their different vulnerabilities, their subjective indicators of well-being were further explored. These included: experiencing severe psychosocial abuse from parents, teachers and police; feeling unsafe because of alcohol and drug abuse; undertaking a range of hard labour including working with mothers to do laundry or cleaning, sorting rubbish at the dumpsites, being forced into prostitution, brewing and selling alcohol, and getting involved in criminal activity. The girls also raised positive aspects of their well-being that they identified with the support they had received in rehabilitation and rein-tegration programmes provided by local organisations such as Pendekezo Letu. These included meeting basic needs and accessing services such as education, health and counselling. They mentioned having food, clothes, shelter, access to sanitation and being in one place so that they could make friends, have time to play and pray, and feel safe, as key dimensions supporting their positive well-being.

On the other hand, the mothers of the girls outlined dreams which included: being able to stop engaging in prostitution; being accepted with HIV and living positively; having a responsible sex life; not using drugs and alcohol; being happy and long-lived; earning money to pay for school fees; improving family income; setting up a small business; and building a house. Many of the mothers said that they were experiencing stigmatisation due to poor health and HIV/AIDS, and that their children had to care for them and work rather than attend school. They wished their children had greater freedom of mobility. They talked about absence of fathers, lack of food and nutrition and limited access to medical services for themselves and their children. In the later phases of the research these subjective indicators of well-being were explored in further depth through examining how intergenerational relationships played out in the context of girls' vulnerabilities.

When mothers and their girls were living and working together and support-ing each other, the team referred to the harm or abuse that children in the family suffered as 'unintentional'. Even where children had to care for their mothers because of AIDS, mothers tended to fully engage in programmes to improve their health and in small loan programmes to generate income. The team made a distinction between this and the 'intentional harm or abuse' that children experienced when parents, often fathers, were perpetrators of physical and/ or sexual abuse and encouraged children into crime, prostitution, drugs and alcoholism. This made a difference to the type of intervention that would be most appropriate for the girls and their families (see further consideration in discussion of results and a new theory of change).

Case studies demonstrating the journeys of girls and their families to the street

This section presents the detailed analysis from six exemplary case studies of street-connected girls. It concentrates on research regarding the girls' journeys to the streets and facilitates further understanding of their situations of living and working on the streets. Social protection interventions that help street-connected girls enter education and sustain interventions to address their poverty and well-being were also examined.

The team analysed the forty-eight case studies from phases 2 and 3 and found that there were some drivers specific to individual informal settlements (e.g. one settlement experienced heavy police, city council, peer and gang violence, while another person cited physical danger from a new road).

Poverty, food insecurity, child labour (including hard labour and prostitution), dysfunctional families (including drug abuse and alcoholism) and sexual and physical violence against children, particularly orphans or step children (from parents, other family members, teachers and other officials) were all features of the lives of girls in most of the areas in the study.

Looking across these informal settlement districts, the teams identified themes that were based around the types of risks that the girls were experiencing with their families and on the streets. Associated with each theme, the researchers identified the individual case numbers of the girls (using the coding system) who were experiencing these risks and identified stories and visual evidence that best demonstrated the girls' experiences of vulnerability and how they coped with different risks. From this analysis a new theory of change has been developed (see Figure 2) on the basis of vulnerability and risk. Six exemplary case studies were selected to show the types of barriers and facilitators that children face in their journeys to the street and to demonstrate the range of social protection interventions that may be appropriate depending on the vulnerabilities and risks that the girls experience.

The case studies have been selected to mirror the selection of detailed case studies overall: two cases for whom the PKL intervention worked; two cases in which the girls had dropped out of school; and two cases where the girls had not yet received any intervention. This selection also demonstrates the importance in the analysis of different family structures and intergenerational power relationships. All cases use the pseudonyms that girls chose during the research.

The first two cases, Mercy and Diamond, show how taking girls to a separate space for rehabilitation and working in parallel with their primary care givers – generally adult female members of their households – on health, counselling and small business enterprises, has worked for the girls and has helped them in their pathways from the street to school, thus realising their right to education.

Mercy's story

Mercy, aged 13, from Mathare, lives with her aunt. She said she wanted to commit suicide after the death of her mother because she loved her very much. After her mother died she was moved around between relatives while working on the streets. She and her family lived in poverty outside Nairobi with little cash to buy food. Mercy moved to Nairobi to live with an uncle and then sold vegetables with her grandmother after school when she was in lower primary. They were able to buy enough food and put money aside for school fees. As her grandmother became ill, she started to work alone while the local bishop helped them with the shopping.

> My parents died and I was taken to live with an uncle. He was not able to take care of me and provide for me ... My grandmother was sick so I had

to help her sell vegetables after 2pm and on Sundays. Unfortunately, she died too. I was adopted by another uncle, who was not able to take care of me. He took me to live with my aunt.

When she and her grandmother sold vegetables they used to sing traditional songs and her grandmother told her stories about her life. Mercy stated that she really misses her grandmother and loved her very much. Now when her uncle isn't able to pay the school fees, her aunt has to borrow from neighbours. Mercy often quarrels with her aunt because her aunt always thinks she's lying. Although they are given money, food and clothes by people in the community, Mercy, her sister and aunt still have to work to get enough money for food. She avoids dark places on the streets where she feels there's more likelihood of abuse. Mercy and her sister avoid the police as they have been accused of being thieves and have been beaten when they couldn't run away (her sister has a bad leg so runs slowly).

Her half-brother Jobe is 15 years old. He tried to live with his father who came to find him, but was beaten and locked in the house: a neighbour had to let him out. Many of his friends are gang members involved in crime. When he tried to jump onto a car to hijack it he fell and had to go to hospital. He was given some clothes by policemen, occasionally had a communal shower which he had to pay for, and went to a mosque for food. On the streets there is a toilet that he used during the day, but had to use the grass at night as the toilet was locked. He was often kicked by 'drunkards' when sleeping on the street, so he tried to shelter in a small shack.

This story of Mercy and her street-connected family demonstrates the complexities facing young girls living in poverty in the slums of Nairobi and how interventions to help girls into education need to be multifaceted to address all of the different risks and vulnerabilities they face. Although Mercy was poor and worked with her grandmother after her mother died, they managed to pay for her to attend school. After having to drop out of school when her grandmother died, she and her aunt met a social worker from Pendekezo Letu who was identifying vulnerable children during outreach work in slum areas. She suggested that Mercy attend their rehabilitation centre for ten months whilst they also worked with her aunt and extended family. As a result, Mercy said that she is more aware of her rights and has learnt basic skills: cleanliness, washing clothes, building on her literacy and sharing her knowledge of rights with others, including her right to education. She is also more aware of abuse from older street-connected girls. For example, Mercy now avoids dark places in the slums and situations where older girls or women have tried to encourage her into prostitution. Her sister was also in the rehabilitation centre and then went back to school, and her brother is in a programme of vocational training to wash and repair cars. Her suggestions about improving support for her education and well-being were: to have a space for guidance and counselling on the streets; to work with parents and guardians to address abuse; and to make sure that girls have freedom of movement and feel safe in

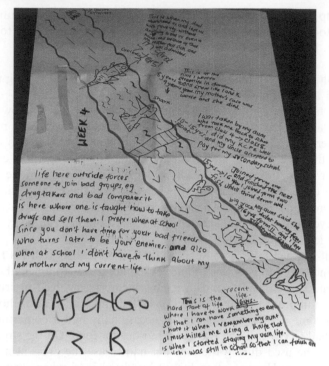

Figure 4.1 An example of a road of life of a street connected girl

different areas on the street. She has subsequently been able to stay at school because Pendekezo Letu have helped her prioritise her education, although she still needs to support herself and contribute to school fees by selling vegetables after school.

Diamond's story

Diamond, aged 15, from Dandora, comes from a family with eight children. Her mother was unable to look after them properly because of poverty and mental health problems. Diamond lived for some time with her grandmother who became bedridden because of AIDS, so all of her children had to work on the streets to try to find enough money for food. Some community members occasionally helped with food, clothes and places to stay. Diamond talked about how she was bullied on the streets and avoided going to parts of the slum where she was sexually harassed by boys. On one occasion, some older boys chased her and friends through the streets and raped her friend. Diamond became sick because of living in dirty conditions on the streets and having nowhere to wash herself or her clothes. She didn't have enough money for food or sanitary towels, which was particularly difficult with no washing facilities. When she was sick she was unable to work to buy food. With friends, she tried to find different places to sleep at night. Her friends started to take drugs and made fun of her because she didn't want to, and she became depressed because she realised she didn't have any true friends.

Pendekezo Letu social workers took Diamond to their rehabilitation centre and worked with her grandmother in their group of women living positively with HIV. In this group they provide food, medicine and counselling to mothers and guardians while girls remain in their care at the rehabilitation centre for ten months. Diamond went back to school but dropped out when she was 13 years old, due to more peer pressure from friends on the streets. Again during the outreach that Pendekezo Letu do in slum areas, she met one of the social workers who encouraged her to go back to school again and she is now still studying. Pendekezo Letu has also helped to find Diamond and her siblings a place to stay at her uncle's house, and her uncle comes to talk to the teacher if any issues arise at school.

The next two case studies, Angel and Maryann, show that when girls end up alone or without support from their families, they experience a number of different risks. The current Pendekezo Letu interventions would need to be built upon in order to provide more intensive support in these circumstances. For example, interventions need to be added or refocused in order to change the behaviour of the adult male members of the household and to address alcoholism, substance abuse and criminal activity by girls and other family members. These cases demonstrate how the safety of girls on the streets and their treatment by other street children and adults in the community present significant risks for girls in their complex lives.

Angel's story

Angel, aged 17, from Thika, was raped on her way home from school. Fortunately, as she struggled to walk home, some neighbours helped her and took her to hospital. Angel's home life was very difficult. Angel said that her mother treated her so harshly that she is frightened of talking to her, even now: in Swahili, 'Mama yangu ni mkali sana ata saa hii namogopa bado'. Following beatings and other bad treatment, Angel ran away from home and started living on the streets. Angel explained how she was introduced into child prostitution by her friend on the street as a means of survival. When her mother was interviewed, she said that Angel ran away from home, got pregnant and was infected with HIV, and the person responsible walked out on her. Unfortunately, after the birth the child died. Angel herself did not talk about rape, being pregnant and her HIV status. Although she went to the rehabilitation centre, Angel dropped out of school and does not want to engage with the programme.

Angel was not only badly abused but, working as a prostitute, infected with HIV and traumatised by the death of her child. She needs different types of coordinated, intensive intervention to have a more sustained pathway out of poverty. This would need effective partnerships between different organisations with the capacity to address the many complex risks facing Angel in her life connected to the streets.

Maryann's story

Maryann, aged 18, from Dandora, lives in the slums with her mum and seven siblings (two boys and five girls). Her parents are divorced and there were problems at home: it was dirty, there was a lack of food, few clothes, nowhere to sleep, and their mum could not take good care of them. She therefore went to the street. She got married at an early age, but is now a single mother. Once on the street, 'senior street children' tried to attack Maryann and beat her and her siblings to chase them away from where they wanted to sleep. They tried to rape her and steal her food. They didn't want her to go anywhere near their territory. Maryann felt that she was stigmatised and hated by people in the community because she was desperate, dirty and had few clothes. During the regular outreach in slum areas by social workers from Pendekezo Letu, one of the team took Maryann to the rehabilitation centre in 2012. She learnt about her rights as a child and she mentions how well she was fed.

After accessing school and spending a year completing Primary 7 following her first attempt at rehabilitation with Pendekezo Letu, she dropped out of school again because of peer pressure from friends who were not at school. She used to take drugs that were given to her by friends and so didn't attend class. Maryann took drugs for one year and then went back to Pendekezo Letu's rehabilitation centre to follow vocational training in hairdressing and tailoring. After getting pregnant, and due to lack of money for enterprise start-up, she is not currently using the skills that she developed during rehabilitation. However, she feels that she at least knows her rights and responsibilities and that in future she could be a good mother. At the rehabilitation centre, said Maryann, she was taught how to respect other people and use good language. She also felt that counselling and information about reproductive health has helped her to think about being a better mother. They not only faced a difficult situation at home, but also an abusive situation on the street. For Maryann, multiple risk factors meant she dropped out of school and that she cannot use the skills she learnt on the vocational training, unless in the future she can obtain funds to help start a small business.

Maryann's story demonstrates that where there are multiple risk factors, there may have to be additional interventions in order for rehabilitation and integration into the community to meet with success in the longer term. When girls are not living and working with their mothers, but end up alone and are facing several risk factors, then they will need to have longer-term and especially targeted interventions, or to be signposted to other services.

The final two cases, of Grace and Mariam, demonstrate how Pendekezo Letu are considering how to address the risks for those street-connected girls who have not previously been involved in any interventions. These cases show the complexity of the girls' lives and how different interventions may be necessary depending on a girl's particular life history.

Grace's story

Grace, aged 19, from Majengo, feels that her mother is completely unreliable and that this has contributed to her dropping out of school. Grace, along with other children from Majengo, goes to the streets in Eastleigh to beg for money and food. Her parents also encouraged this behaviour. Grace says she is forced to take drugs and tobacco: 'tulilazimiswa kuvuta bangi na tobacco'.

She was introduced to drugs by friends while she was on the streets. The older street boys and girls used to force the young street children to smoke bhang and sniff glue. When interviewed, her brother said:

> Life here outside forces someone to join bad groups e.g. drug takers and bad company. It is here where one is taught how to take drugs and sell them.

Grace dropped out of school because it was hard to afford school fees. She also felt stigmatised as she was 'slow' at learning, and was older than the other children. She feels that social branding is pervasive in Majengo slum and many street-connected children complain that at some point in their lives someone in the community has labelled them as a 'chokoraa' (scavenger).

Grace has been chosen as an exemplary case as she faces multiple risks and is therefore regarded by Pendekezo Letu as extremely vulnerable. Where both parents and girls are taking drugs and alcohol, additional interventions may need to be developed to deal with the complex and multifaceted vulnerabilities and risks that girls experience in their lives. Grace's mother is an addict of 'illicit brew' (locally made alcohol). Pendekezo Letu recognises that for girls facing multiple risks, the intensive rehabilitation over a short period of time at their centre may not be enough to prevent dropping out of school. They have therefore started to pilot interventions working with those experiencing alcoholism and substance abuse, and are also considering how to work in partnership and to refer girls and their families to other organisations with expertise on these issues.

Mariam's story

Mariam, aged 13, from Korogocho, was born after a man raped her mother. Her mother obtained a job as a home help (domestic servant) with a family that treated her well, but the head of the household raped her and this left her deeply scarred. She became pregnant and gave birth to a daughter – Mariam. She then found a partner and had another daughter with him, but he ended up abusing Mariam and then chased her from the house when her mother was away. Her mother left him to become a single mother of two. Mariam then had to drop out of school when she was in Class 5 due to lack of finances. Mariam and her sibling began accompanying their mother to the dumpsite to scavenge for food, scrap metal and plastic bags. Mariam now washes clothes for other people in order to earn a living, in return for either food or cash, while her younger sister continues to sort rubbish at the dumpsite with their mother. Despite Miriam's labour and abuse, the Pendekezo Letu intervention may well be suitable for helping Mariam and her family get back on their feet. With rehabilitation for Mariam and her sister, and support for their mother to set up an enterprise, they may be able to return to school and sustain their education with income. Pendekezo Letu's short-term intensive rehabilitation has been found to be particularly successful in these circumstances: girls living and working with their mothers or other members of the extended family with whom they have good relationships (Johnson et al. 2015).

A new theory of change for interventions with street-connected girls

A greater understanding of the subjective indicators of well-being and the vulnerability of the street-connected girls participating in the study have helped Pendekezo Letu to understand the most appropriate interventions that can be provided to help street-connected girls to realise their education and living rights. This child-centred research has informed a new theory of change that builds on the existing Pendekezo Letu programme. For instance, it supports the current approach to individual empowerment of girls in changing their attitudes and behaviour, while building on their knowledge of rights and their literacy. It also acknowledges the importance of economic strengthening through support in vocational training for older girls and in small-scale enterprises for primary care givers. This support for girls and their mothers seems to have been most successful where girls are street-connected while retaining some support from their care givers, usually adult female members of their families – mothers, grandmothers and aunts. The analysis of the perceptions of girls showed the kind of intervention that had helped them to deal with their vulnerabilities and the risks that they faced. Where girls face multiple risks in their complex lives, additional interventions are needed, especially where girls are alone or in particularly dysfunctional families. Situations of abuse and neglect had often led them to the streets. This theory of change also takes into account how working with extended families and communities is necessary in order to break intergenerational transmissions of psychosocial aspects of well-being and poverty (as suggested by Moncrieffe 2009). The ethnographic and participatory approaches used can also help to illuminate girls' experiences of place and mobility (as also suggested as important by Porter and Mawdsley 2008) and to understand both their agency and vulnerabilities (as suggested by Mizen and Ofosu-kusi 2013).

All of the street-connected girls with whom the research team worked are 'marginalised' due to a deprivation of rights: they are all in a situation of poverty and living in informal settlements in Nairobi. They are all also participating in child labour and have a lack of access to education due to their situation of poverty, the gender discrimination they face, and the work that they have to do to survive. Through the participatory child-focused research, the team found that street-connected girls, despite all being 'marginalised', experience and are further affected by different psychosocial and emotional factors. Their complex experiences may therefore be further unpacked in terms of their vulnerability and exposure to different types of risks. By looking across the research locations and case studies, including considering the different categories of difference relating to the girls' identity such as their age, education, family and living situation, a theory of change has been developed that is based on a spectrum of vulnerability. This can help Pendekezo Letu and other organisations working to get street-connected girls to meaningfully engage in education and to help their primary care givers to support them and sustain their pathways out of poverty. It is intended that this theory of change can help Pendekezo Letu to plan interventions that are appropriate to the complex lives of street-connected girls and their families in Nairobi.

The vulnerability of girls in this theory of change is determined by the different risk factors that the girls experience. Associated with these are the types of interventions that girls felt most addressed their needs and helped them to realise their rights, including their rights to education and feeling a sense of safety and belonging in their schools or educational settings and living situations. Interventions involve working with the street-connected girls, but also with their families, communities and local policy makers.

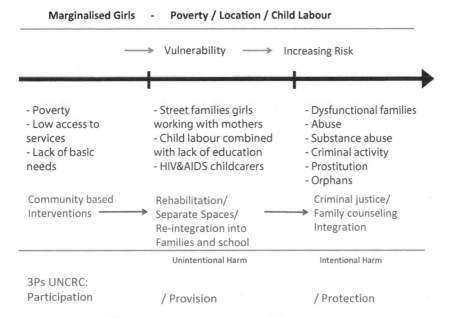

Figure 4.2 Vulnerability spectrum: a new theory of change

At one end of the spectrum are girls who are marginalised and live in poverty, but are less vulnerable as they have support from their family, a strong group of peers or are already supported by other organisations. They suffer from inability to access services, and thereby they lack basic needs and cannot fully enjoy their rights. These girls also experience poor sanitation and living conditions, and need to work to contribute to household income. The research showed that girls in this situation could benefit from community-based interventions with a focus on awareness of rights and access to services. Examples of successful interventions included helpdesks based in the community to provide street-connected children with advice on child rights and gender violence and health camps – in effect providing girls with help on the move. Pendekezo Letu has also started to run a programme where they provide free sanitary towels for street-connected girls and raise awareness of this throughout different communities. Churches and religious centres, as well as community resource centres, have been shown in some of the case studies and from reference group discussions to fulfil this role of providing advice and acting as community-based referral systems to other more specialist services.

In keeping with this community-based model, Pendekezo Letu has started to pilot a Community Based Child Protection Committee in Korogocho. This includes community members, the chief, local police, and the government voluntary children's officer. They organise some minimal interventions for rehabilitation of street-connected children such as guidance and counselling, and mainly act as a referral body to provide information for children and parents who want to go to school or access health services. They also talk with community members to negotiate conflict amongst adults or conflict between adults and children, acting as mediators to resolve situations that may otherwise lead to further abuse or discrimination of the street-connected children.

By contrast, girls in the middle of the vulnerability spectrum are often living and working with primary care givers, usually their mothers, grandmothers or aunts. As demonstrated in the case studies, their fathers are often absent or abusive. These girls experienced what the research team called 'unintentional harm'. That is, physical and psychological harm that was caused due to poverty: families living in inadequate shelter with a lack of access to services to meet their basic needs, including a lack of basic education. Mothers, and sometimes children, were affected by HIV, and girls often acted as carers for their parents. In order to survive children had to work: where mothers were well enough children worked with them, although sometimes girls had to go out by themselves. They were involved in a range of different, sometimes abusive, child labour activities, including sorting through rubbish at the dump, street retail, and prostitution. Therefore, where girls were vulnerable, but living and working with family members and experiencing one or two risk factors in their lives, the Pendekezo Letu rehabilitation programme was more able to transform their lives and sustain routes into education: the girls responded well to counselling and improved their behaviour and literacy, and ultimately were able to attend and stay at school.

The rehabilitation programme run by Pendekezo Letu involves taking vulnerable girls to a separate rehabilitation space in Thika for a set period of time (ten months) and working with them intensely on behavioural change, literacy and numeracy as well as building their confidence and self-esteem. This separate space, although requiring relatively high funding over the ten-month period, is considered to be necessary because of the risks that the girls face in their everyday lives. Indeed, they require a space for rehabilitation away from the difficult situations on the street because achieving sustainable and rehabilitation using participatory approaches is far more challenging within the girls' usual community or family contexts. This approach also fits with the idea of creating safe spaces off the street where girls can be supported in addressing psychosocial aspects of their well-being, including confronting the intergenerational power relationships that they may be going back into. The approach is therefore linked with the concept of spaces for participation and understanding the 'intergenerational performance' that affects the agency of children (suggested by Mannion 2010). This strategy also addresses the three Ps of the UN Convention on the Rights of the Child (the UNCRC): Protection, Provision of services and Participation. The girls need both protection and basic provision of services in terms of sanitation, food, shelter and education, in order to feel able to participate in changing their own behaviour, their awareness of their rights, their attitudes and their aspirations.

In addition, siblings attended training and received guidance for income generation, for example in hairdressing and motor mechanics. At the same time mothers or other primary care givers met in groups to discuss healthcare, including living positively with HIV, and were helped to provide their family with food until they could take advantage of small loans to build up local enterprises. Once the children returned to school, small clubs were set up to support their integration into school, while families continued to attend family counselling. The combination of interventions provided by Pendekezo Letu were successful for those girls who were living with street families and/or working with their mothers or other primary care givers, as well as for orphans who retained at least some extended family support. At the household level these interventions introduce rights-based approaches to family members, but also address the corresponding 'responsibilities', which is an aspect often overlooked in development interventions at the local level.

A detailed analysis of the vulnerabilities and risks for girls, including their interactions and support from their families and adults in the community, is required to understand how the intergenerational transmission of poverty can be prevented. Alongside the direct support provided to street-connected girls and siblings, Pendekezo Letu works with their primary care givers on HIV care and small loans for businesses. They are trained in 'asset based community development' and business skills, and then provided with modest cash grants of Kshs 5,000–10,000 to start up small enterprises. Social workers follow up on the progress of these mothers, and they are encouraged to form 'saving and internal lending communities' (SILCs) where they meet regularly and promote each other. Grants are mainly provided to enable families to start the businesses and repayment is not required until the parent or other carer is able to provide food and shelter for their family and afford fees for their school-going children. Consequently, mutually reinforcing connections are established between social protection mechanisms and child protection systems.

Finally, the right hand side of the spectrum represents the most vulnerable girls facing multiple risks, who live by themselves on the streets or in highly dysfunctional and abusive families. The abusive situations that girls experienced are what the research team referred to as 'intentional harm': girls suffered from violence in the family, at school or in the workplace, and were subjected to a combination of different high risk situations. These risks included alcoholism and substance abuse in the family; girls and siblings being forced into abusive situations of child labour, including prostitution; drug dealing or selling illicit brew or alcohol and other criminal activities; and continual referral between family members and/or carers. Towards this end of the vulnerability spectrum more psychosocial and emotional aspects of the girls' well-being need to be addressed, as well as the abuse and harm caused by dysfunctional families, in order to break intergenerational transmissions of poverty.

These highly vulnerable girls were not able to sustain the benefits of the current Pendekezo Letu intervention described above and dropped out of the education system even if they had been able to enrol and attend for a short period of time. In response therefore, other additional forms of intervention are needed for these girls and their families. These include negotiating increased access to support in the criminal justice system, additional emphasis on providing start-up training as appropriate alongside training in small enterprise development for girls, their siblings and parents, and additional work and counselling on alcoholism and substance abuse within families (see the Conclusion below). Interventions need to specifically focus on abuse from fathers when girls are suffering from highly abusive situations. On occasions it was also the case that officials and police were themselves violent towards street-connected girls. Consequently, there is a need for additional training for these services. In one or two cases, school teachers were abusive, suggesting a need for further training on alternative forms of punishment and on preventing discrimination against street-connected girls. Pendekezo Letu have been working to support girls, their siblings and other young people in the community who are caught up in the criminal justice system by offering legal support when they are in conflict with the law.

In order to fully understand interventions to improve the well-being of street-connected girls, Pendekezo Letu has now adopted a more participatory approach to understanding girls' experiences of risk and vulnerability in the context of the inter- and intra-generational power dynamics they face in informal settlements where they live and work, as well in their mobilities onto the street and from street to school. The research has therefore been called *Mitanni Hadi Shuleni* ('From Street into the School') by the local Kenyan team.

Conclusion

This robust, child-centred, qualitative research with over 200 street-connected girls has directly informed the way in which the Kenyan NGO Pendekezo Letu works with street-connected girls in the slums of Nairobi. It presents original evidence on the journeys of marginalised girls to the streets and the situation that they find themselves in when they live and work in the slums or informal settlement areas of Nairobi. Intergenerational relationships affect the risks girls face in their lives, but so also do the successes of interventions aimed at improving their access to education and health services, and the realisation of their living rights (as defined by Hanson and Niewenhuys 2013). The research also offers new insights into how vulnerabilities of girls and their families can be locally constructed, through understanding their intergenerational relationships and their subjective indicators of psychosocial well-being. These include their experiences of place and mobility. This analysis in turn informs a variety of social protection and child protection interventions for street-connected girls and their families who are living and working in marginalised situations. It is undoubtedly the case, however, that more research is required in order to further understand the vulnerabilities, agency and complexity of the lives of street-connected girls and their families in Nairobi and other informal settlements in Africa and the global South, and to apply this to interventions and policy.

A core element of this qualitative child-centred process of research was to have a well-developed safety and ethical framework and set of procedures to support both the researchers and participants in the research process. A range of national and international decision makers and policy makers were engaged in reference group discussions through international webinars (as part of the UNGEI programme), as well as through national research dissemination and uptake events in order to enable national and international communication. As a result, lessons learned on theorising vulnerability and agency when working with street-connected girls and their families is progressing and more lasting change can be achieved in the longer term. While inter-agency and inter-disciplinary activities to mobilise these approaches at national level are improving, they require further formalisation and support to remain sustainable and connected to global best practice.

Ultimately, if interventions to improve the well-being of street-connected girls in Nairobi help us gain a fuller understanding of children's experiences of place and mobility, risk and vulnerability, and span generations, intergenerational transmissions of poverty are more likely to be disrupted or broken. The girls' own definitions of psychosocial aspects of well-being, including the root causes of their mobilities to the street, their journeys to the street and to school are a critical part of this process. The ways in which inter- and intragenerational relationships support or hinder them in these journeys should also be considered. This approach recognises that economic strengthening alone (at the individual or household level) is insufficient as a response to marginalised street-connected girls. Without a holistic approach to improving well-being, the causal factors that lead girls to the streets, especially those girls from dysfunctional families, may not be fully addressed, thus hampering the sustainability of their pathways out of poverty and/or their routes into education.

In terms of methodological responses to this situation, it is important to recognise that the mobilities of street-connected girls and the spaces they occupy on the streets are subjective and dynamic, constituted by social and power relationships in families and communities. This suggests the need for more qualitative research on place, space and mobilities that respects children as active participants in order to understand the experiences and journeys of street-connected girls and boys. The participatory research presented in this chapter used ethnographic and visual methods such as roads and rivers of life. Other studies support the importance of understanding children's mobilities in sub-Saharan Africa through ethnographic and participatory approaches (as also used by Porter et al. 2010).

Using these principles, a capacity building approach was taken in the research in Nairobi, in which the team co-constructed the conceptual framework and planned the detailed research. Subsequently, Pendekezo Letu has started to apply creative methods for identifying children and adults as suitable to participate in their programmes and to establish rigorous, ongoing participatory evaluation and impact assessment.

Beyond these approaches, Pendekezo Letu continues to follow up with local and national policy and strategies for working with street-connected girls. This is a key role for them, as they are well positioned to innovate with tailored approaches in addressing intergenerational transmissions of poverty by working with marginalised street-connected girls and their 'street families'. They are also well placed to communicate findings to national-level actors. These approaches demonstrate that an understanding of the synergies between child protection and social protection are fundamental to addressing the needs of street-connected girls in both the short and longer term. It also highlights the importance of children's participation in understanding subjective indicators of psychosocial aspects of well-being and how girls experience mobilities and place, risk and vulnerability.

Acknowledgements

The team would like to thank all of the street-connected girls, family members, community members, and local and national stakeholders working to provide services for street children. The full research team included the following: Dr Vicky Johnson, research lead; Laura Johnson, research coordinator; Boniface Okari Magati, national research coordinator; David Walker and Allan Kiwanaku, research advisors; the following researchers/social workers from Pendekezo Letu – Sarah Wanjiru Mbira, Magdalene Waithera Munuku, Esther Waithira Mwangi, Jane Gituthu Muthoni, Winnie Wanjiku Wanjiru, Job Ndirangu Nduhiu, Florence Koki Mwania; and the following young researchers – Abel

Mayieka Dennis, Owour Victor Otieno, Mbatia Serah Faith Wanjiru, Warugu Purity Wanjiru, Mary Njeri King'ara, Silvester Kimari, Moses Ndung'u Chege, Brian Wanene Waiyaki, Waweru Joanne Njoki and Irene Ndulu Mully. The research team showed incredible determination and dedication in working with street-connected girls in order to understand the complexity of their lives and the interventions that would help to improve their education and well-being.

5 Tales of the map of my *mobile* life

Intergenerational computer-mediated communication between older people and fieldworkers in their early adulthood

Sergio Sayago, Valeria Righi, Susan Möller Ferreira, Andrea Rosales and Josep Blat

Introduction

'Mobilities research thinks about a variety of things that move including humans, ideas and objects. It is particularly interested in how these things move in interconnected ways and how one may enable or hinder another' (Cresswell 2011, 552). Nowadays, it is increasingly common for members of the same family to be living in different regions, countries, or even continents to those where their grandparents live. This physical distance, compounded with the intensity of current life in western countries, hinders intergenerational communication within families. However, Computer-Mediated Communication (CMC) opens up new opportunities, as it 'spans space and time barriers allowing a person to work, learn and communicate from those times that are most convenient for him or her' (Zaphiris and Sarwar 2006, 404). How is the 'mobility turn' (Sheller and Urry 2006; Hannam et al. 2006) addressed in CMC research with older people (60+), and what can studies conducted in this field offer to the understanding of mobility and ageing? This chapter addresses both questions by building upon a literature review and a decade of fieldwork conducted by the authors in three European cities (Barcelona and Madrid in Spain, Dundee in Scotland) with approximately 700 people aged 60–90 with different levels of previous experience with Information and Communication Technologies (ICT). This chapter presents tales of an intergenerational CMC wherein older people's *mobile* lives, for example through migration, changes in their vital landscape and commuting, played a pivotal role. Ordinary ICTs not specifically designed for older people, for example Google Maps, were key means of enabling intergenerational CMC with early adult fieldworkers,[1] wherein older adults were the protagonists and shared their life experiences, rather than family matters, which predominate in a great deal of intergenerational CMC research conducted with older people.

Related work

Whilst older people and ICTs might be two worlds apart, a great deal of CMC research has been conducted with people aged 60+. This might be accounted for by the importance of communication in ageing (Nussbaum et al. 2000) and the mobility of younger generations within extended families. We seek to provide an overview of the research conducted with older people in CMC and discuss it through a mobility lens in an attempt to strengthen its connection with the theme of the book. We do so by reviewing studies we regard as highly representative of the CMC research conducted with older people since 2000 and strongly related to this chapter.

CMC in virtual/online communities

A number of studies have explored older adults' attitudes to, and how they either actually use or would use, cyberspace (Blit-Cohen 2004), virtual/online communities (Lepa and Tatnall 2002; Kanayama 2003; Xie 2008; Pfeil et al. 2009), public online newsgroups (Zaphiris 2006) and social networking sites (SNSs) (Gibson et al. 2010; Righi et al. 2012; Ferreira et al. 2014). This research tends to highlight that online/virtual communities can potentially be useful for older people to keep in touch with peers interested in the same topics as they are. Online communities can present older people living in remote areas, or those who suffer from a specific illness or are housebound, with an opportunity to meet people who are in a similar situation and engage in satisfactory social interaction (Kanayama 2003; Xie 2008; Pfeil et al. 2009; Zaphiris and Sarwar 2006). Also, and despite being widely regarded as a very heterogeneous user group (Gregor and Newell 2001), this research shows that older people with different cultural backgrounds voice remarkably similar privacy concerns in their use of SNSs, for example being reluctant to meet unknown people online, make their own photos or videos public to the 'whole internet', or reveal personal information (Gibson et al. 2010; Righi et al. 2012; Ferreira et al. 2014).

Intergenerational CMC within extended families

The 'modern western way of life often leads to a geographical separation of grandchildren and grandparents' (Fuchsberger et al. 2011, 50). Hence, a number of studies have focused on understanding how older adults and members of their families communicate with existing CMC tools. In Sayago and Blat 2010, approximately 350 older people adapted to their interlocutors as they wanted to be socially included. They used video chats (for example Skype) rather than e-mail to keep in touch with their grandchildren (aged 5–9) on a regular weekly or fortnightly basis in 20–30 minute sessions, as e-mailing was a much bigger effort for their young grandchildren, with low literacy, than chatting. However, the same group of older people both preferred and used e-mail to communicate with their grandchildren aged 10+ (and adult children too), because of their hectic work, study and social agendas. In Dickinson and Hill (2007), e-mail was often initially adopted by nine grandparents, who lived in their own homes, as a way of staying in touch with their grandchildren, as e-mail allows asynchronous communication irrespective of time differences and enabled them (and their grandchildren) to communicate when they felt like it, as opposed to telephone calls. Older people, despite well-known difficulties in intergenerational communication practices due to busy schedules or extended family members' lack of technology use[2] (Tee et al. 2009), tend to find this type of intergenerational CMC 'worthy of time and dedication' (Lindley et al. 2009).

Other studies, larger in number, have designed new systems to connect aged parents and grandparents to their children or grandchildren over a distance through the sharing of location-based video stories (Bentley et al. 2011), media gifts (Kim et al. 2013), photos and calendar information (Brush et al. 2008; Lindley et al. 2009), or by means of multifamily media spaces (Judge et al. 2011), social media (Muñoz et al. 2013; Cornejo et al. 2013) and intergenerational play (Feltham et al. 2007; Vetere et al. 2009; Chua et al. 2013). The content of the intergenerational communication most often described when using already existing or more novel CMC tools is of memories of family events, daily, leisure and amusing activities (such as grandchildren playing in the garden) and social support exchange. More recently, the use of audio-enhanced photos to support communication between institutionalised older people with aphasia and family members has been explored (Piper et al. 2014).

How is the turn to mobility reflected in this CMC research?

A lack of physical mobility or difficulties in 'getting out' predominate in the discourse of the potential usefulness of virtual/online communities for older people who are either living in rural or remote areas or are homebound (Kanayama 2003; Xie 2008; Pfeil et al. 2009). The contact initiators of intergenerational CMC within the same family are predominantly the younger generations (Kim et al. 2013). This might be accounted for by the fact that younger generations are either digital natives (Prensky 2001) or more active ICT users than their parents or grandparents, and also that older people are bounded in space, unlike their children and grandchildren. Thus, one could conceive of older people as being unable to use contemporary ICTs due to age-related changes in functional abilities and little (or a lack of) experience with ICT, or as using them in 'extraordinary' ways.[3] Designing new CMC tools in a way that makes them more accessible to older people is to be commended. Also, providing younger generations with novel and interesting ways of keeping in touch with their older relatives might be a way to foster and enrich intergenerational CMC within families. Yet these new developments fail to answer the question of whether older people can become the initiators of intergenerational conversations with contemporary CMC tools and, if so, how these encounters materialise. The tales of intergenerational CMC presented in this chapter answer this question.

Over a decade of fieldwork

We are Human-Computer Interaction (HCI) scholars whose aim is to reveal and explain how older people interact with and make use of contemporary ICT in out-of-laboratory conditions in order to inform the design of more accessible, easy-to-use and meaningful ICTs in their everyday lives. We consider that depth, natural settings, intensity, holism, non-judgmental orientations and giving voice to people in their own local contexts (Blomberg et al. 2003; Hammersley 2007; Fetterman 2010), which are foundational elements of ethnography and participant observation, should (and could) help us achieve our objective. Since 2005, we have been conducting ethnographical and participant observation studies in which a number of older people motivated to discover, learn and become independent ICT users have participated actively. They had very different levels of practical knowledge of ICT, ranging from those who had never used computers and the internet before to those who owned smartphones and had been using computers for more than twenty years. The fieldwork activities encompassed: three long-term ethnographical studies (2005–2008; 2010–2013; 2011–2015) carried out at a 30-year-old adult educational centre called Àgora (AG), in Barcelona (Spain); an 18-month ethnographical study in the Dundee User Centre (DUC), a drop-in clubhouse physically situated within, and run by, the School of Computing at the University of Dundee (Scotland); and two participant observational studies (February–May 2013; October–December 2013) in Espacio Caixa Madrid (ECM), one of the centres for older people owned by the Spanish savings bank foundation, Obra Social 'la Caixa', in Madrid, Spain.

Approximately 700 people aged 60–90 (hereinafter, participants), took part in these studies. We observed and talked to them through informal face-to-face conversations, semi-structured interviews, workshops, hands-on sessions and focus groups. Our observations and conversations took place in different scenarios of situated ICT use, for example e-mailing their grandchildren and friends, setting up (and deleting) their accounts on Facebook, playing casual games and making digital videos with tablet computers and camcorders. These activities took place in courses, workshops and drop-in sessions[4] either run or supported by the authors in the different settings. In these activities, participants played an active role in, for instance, deciding which ICTs they wanted to use and discussing their reasons for incorporating them, or not, in their everyday lives.

In keeping with long-established data gathering practices in ethnography and participant observation, we wrote down most of our first-hand observations of, and conversations with, the participants while taking part in the afore-mentioned activities, or immediately after them if the activity was so dynamic that it hindered simultaneous note-taking. These notes were taken either on paper or on a PC/laptop. Data analysis combined grounded theory (Charmaz 2006) and thematic analysis (Braun and Clarke 2006). The ethnographical and participant observation studies are detailed elsewhere (Ferreira et al. 2014; Sayago et al. 2013; Righi et al. 2012; Rosales et al. 2012; Sayago and Blat 2010).

Given that not all of us were involved in all the fieldwork activities, we put together our fieldnotes and experiences of being in the field in the results, which are presented in the form of realistic and confessional tales. As defined in van Mannen (2011, 48), realistic tales 'focus on minute, precious, mundane details of everyday life', while confessional tales concentrate on 'showing how the technique is practiced in the field' (ibid., 73–74). This combination of styles attempts, first, to show what the participants did and said and what it meant to them, second to give voice to both the participants and the fieldworker in order to understand the impact of the latter on the findings, and third to show how the mobility of older people emerged in these intergenerational CMC encounters *in situ*. We edited the tales in a way that made them representative of the most common types of intergenerational CMC that took place between older people and the different fieldworkers at different times. We witnessed two main types of CMC. One is strongly related to the intergenerational CMC that predominates in previous research. Examples are participants who are grandparents turning to video-conferencing systems in order to keep in touch with young grandchildren whose writing skills are not good enough to establish fluent written communication (Sayago et al. 2011); and older people sharing YouTube videos in a private and meaningful way with close friends, children and grandchildren (Ferreira et al. 2014). The other type, on which the three tales presented in this chapter focus, reveals a different kind of intergenerational CMC.

Tales of the maps of older people's mobile lives

The following three tales were edited and created by translating selected parts of our fieldnotes into English[5] and adding details of the authors' lived fieldwork experiences that connected and enriched them.

Tale 1 – Dundee when I was young

'*Fit like* [How are you], Peter?' I asked one of the participants. '*Nae bad the noo* [so far so good],' he answered. 'Have you seen these pictures? They're *aboot* [about] the old Dundee. I can tell you what they depict, if you want.' 'That would be terrific,' I answered, while sitting next to him and looking at black-and-white pictures on a website he had opened on his PC. Peter started to talk to me about how streets and buildings of the city portrayed in the pictures had changed over time: 'Do you recognise this street? This is [name of the street], when now we have most of the shops in Dundee. And this is the famous bridge [...] Did you know that the bridge collapsed and then it was built again?' he asked me. 'Did what?' I exclaimed in surprise. 'Oh aye, I still remember what happened. It was chaos, a bit of a disgrace for the city. People were quite proud of the bridge, you *ken* [know]. There is an expression local people use that makes reference to that event.' When Peter was about to tell me what these words were, the chairman of the DUC, who was making tea, jumped into the conversation: 'Don't say it, Peter! He needs to figure it out on his own. It's much funnier and a sort of a challenge [smiling]! I've got more memories of the bridge, but it's time for our coffee/tea break. We'll talk about it then, shall we?' From that day on, it was rare to have a drop-in session in which no participant shared with me, mostly face-to-face, but also in short e-mails [two or three lines at most] their memories and experiences of living in old Dundee. They were the protagonists of the tales. When the drop-in sessions were attended by three or four participants, all I did was to sit, listen and learn from their reminiscences, which touched upon important places such as the local area where they grew up, and how these places had either changed (with new houses, shops and department stores) or stood the test of time. These tales were primarily supported by websites that participants already knew and digital pictures they kept or sought online. Participants also brought books and old newspaper clippings, that they kept at home, to the DUC for '*ye tae* learn more *aboot* Dundee', as they said to me.

Tale 2 – The story of my life before I forget

'*Bon dia* [Good morning] Sergei, this your name in Russian,' Rosa, an AG participant, said to me before a hands-on session on e-mail was due to begin. 'Oh, I'm impressed. What is your relationship with Russia?' I asked her. 'I grew up and got married in Russia.' I wanted to know more about this aspect of her life, but I was running the session. The next day, while I was checking my e-mail in the internet room at AG, Rosa showed up and sat next to me. 'Hi, are you still impressed by my Russian?' she asked me. 'Absolutely' I answered. Rosa started to talk to me while typing words in Google:

> I had to leave Spain when I was a very young kid. I couldn't stay in the country, because of Franco's dictatorship. I guess you might know the reason. I was a refugee in Russia. I grew up in a school for girls. I still remember the harsh winters, the language barriers and isolation of the first weeks until I started to speak some Russian [...] Months and years went by, and I met my husband in a very nice square in the city centre. I got married and came back to Spain. Why am I telling you all these things? I want to tell you that I have been diagnosed with having Alzheimer's disease, and before I forget, or I stop taking your courses, I want to share this slice of my life with you.

I was speechless. Before I could say anything, Rosa gave me a virtual tour of the city centre of Moscow and towns nearby by using Google Images. This tour lasted around 2 hours. While she was talking to me, she was able to find photos of the girls' school she attended, parks, streets and public buildings online. Rosa shared stories of her youth at that school, her memories of playing and dancing in the parks, and how, and where, she met her husband. I asked her numerous questions. She answered almost all of them with tears in her eyes. 'Do you understand now why I can't write about all these things in e-mails or these webpages in which everybody can read what others have written?' Rosa asked me. 'These things mean a lot to me' she said. 'Yes, I see now why' I answered.

Tale 3 – My hometown and local area

All participants (at the AG, ECM and the DUC) were interested in Google Maps. They reported that they had seen other people using it and talking about the possibility of seeing their houses, cars and other parts of the cities on the computer screen. In informal conversations, especially when we were starting to know our participants better, the topic of 'I'm (not) from here' was recurrent. Hence, we decided to ask participants to show us their hometowns on Google Maps, as this could be a meaningful way for them to use this technology. This exercise turned out to be exciting. The main reason for this excitement, at least amongst the AG participants, was that most of them had to leave their hometowns when they were children. As Paco put it:

> Most of us are immigrants. We're born in other regions of Spain. We had to leave our hometown when we were children in order to have a future. I've been living in Barcelona for at least 40 years, but I'll never forget my origins [...] Look! I've found my hometown. I can see the river where my mum washed the clothes. Oh, nice memories ...

In courses, workshops and drop-in sessions, participants located their hometown on Google Maps and shared with us their memories of having lived there. These face-to-face conversations were so rich that whole sessions were sometimes devoted to remembering their hometowns in this way. In these sessions, participants would locate the house in which they were born, or the river where they fished, or how they made do with almost no money.

These conversations were digital, too. Indeed, it was not uncommon for the participants to e-mail us, and to other participants as well, screenshots of Google Maps inserted into Word or PowerPoint documents, in which they shared their memories in a highly visual way. Participants at the AG, ECM and DUC also created digital videos with the screenshots by using Windows Movie Maker and sent them to us via e-mail. These e-mails triggered both face-to-face and online (e-mail) conversations, which, on the one hand, helped us to greatly improve our understanding of Spanish, Catalan and Scottish geography and culture but, on the other hand, altered our plans for some fieldwork activities, which we had to postpone or change as the study progressed, because they did want to share so much of their *digital* memories.

A different topic of these mobile 'Google Maps-mediated-communications' was the participants' local area or neighbourhood. Participants with more practical knowledge of ICT were the leaders of these conversations. While heading to the metro station near ECM, one of us came across Maria, who had to leave the session. 'Oh, hi, sorry I had to run away, but I had an appointment with the doctor [...] By the way, let me show you something on my tablet,' she told me. 'What is it, Maria?' I asked. While tapping into the Google Maps apps, she told me:

> It's only a minute [waiting for her tablet to launch Google Maps]. Now, see, we're here. Do you remember you asked me where to eat good Spanish soup? You must go to this restaurant, which is very near the metro station where you get your train back home, you see? There you'll eat the best consommé in Madrid. I'll e-mail you the directions if you want, or show you where it is now, if you want, since it's on our way to the metro station. I'm paying a visit to a friend of mine and I'll take the metro [looking at Google Maps again] in the Bilbao station, which is where you're heading to [smiling]!

Discussion

As stated in Cresswell (2011), mobilities research is interested in how humans, ideas and objects move and in their interrelationships. How is this mobility addressed in the results? In Tale 1, the extent to which objects, particularly architectural landmarks of the cities in which older people have lived most of their lives, have evolved or stood the test of time, played a pivotal role in intergenerational conversations, which were supported by digital and non-digital artefacts and fuelled by the desire of older people to share their knowledge of a city with us. Tales 2 and 3 reveal snippets of older adults' *mobile* lives. Tale 2 shows how the life course (childhood and youth) of an older person, along with the historical context that shaped it, and current changes in mental health, triggered intense and emotional face-to-face conversations mediated by, sometimes triggered by, CMC tools. The virtual tour of places and land-marks in Tale 2 are extended in Tale 3, wherein the geographical mobility of older people, reflected in migration, commuting and knowledge of a local area, was the cornerstone of intergenerational conversations, which some-times hindered already planned fieldwork activities and were on the move too, enabled by mobile technologies. This intergenerational CMC on the move might be surprising, as it does not seem to correspond with mobility scenarios most of us are familiar with, such as middle-aged business people talking on their phones while commuting.

Mobilities and intergenerational CMC

The mobility paradigm has proven useful to understand further a number of practices, spaces and people (Cresswell 2011). This chapter suggests that

CMC studies that take the mobility of older people as the central fact can extend the field of mobilities. The tales show how our participants took their time to share their memories and life experiences with early adult fieldworkers, sometimes with great passion. While older people might not keep in touch with unknown people online (or even face-to-face), for our participants, engaging in intergenerational CMC with us – people with whom, despite not being family members, they had established a good and long-term relationship – was 'worthy of time and dedication' (Lindley et al. 2009). They were the protagonists of the tales, which were triggered by their mobile lives through migration, changes in their vital landscape, and commuting. The participants and their tales were enabled and supported by a number of ICTs. This challenges stereotyped views of older people when it comes to ICTs (Durick et al. 2013). The personal, private and intimate elements of some of these conversations were key reasons in accounting for a strong preference for one-to-one conversations, either face-to-face or online, mediated by digital pictures and web maps, instead of SNSs. This result reinforces privacy concerns shown by older people in relation to SNSs (Gibson et al. 2010; Righi et al. 2012).

The intergenerational CMC presented in this chapter is also relevant for research into ageing. First, across disciplines devoted to the study of ageing such as psychology, sociology and biology, 'the idea of ageing as a lifelong process [...] of cumulative advantage and disadvantage [...] appears to be universal' (Gans et al. 2009, 729). In other words, it is difficult to understand an older person if we do not consider his or her past and current aspirations. Yet, it is striking to note that this common way of understanding ageing is not clear-cut in the content of much of the intergenerational CMC research reviewed in this chapter. Second, in *The Futures of Old Age*, Vincent et al. (2006) predicted that migration would become an important issue in the likely futures of old age over the next 30 years, and that this migration should be 'understood in [its] household, family and temporal context' (ibid., 217). This chapter shows that this predicted future is already here, at least in light of the importance of migration in Tales 2 and 3. Third, the three tales challenge intergenerational communication practices, which are heavily dominated by over- and under-accommodation (Giles and Gasiorek 2011): when people interact they emphasise or minimise the social differences between themselves and their interlocutors through verbal and nonverbal communication. While these accommodations may be natural reactions when first encountering older people, our decade of fieldwork is mostly composed of stimulating and emotional conversations.

Strong and weak points

The long-term aspect of the fieldwork activities, the large number of older people who participated in them and their cultural diversity, might be the strongest elements of the research presented in this chapter. However, the profile of the participants, who were interested in ICTs and sharing their life experiences with us; the settings, which encouraged socialisation and learning; and our role as fieldworkers interested in learning with and from our participants, are likely to be the main limitations of the results. Further studies are warranted in order to corroborate or challenge the intergenerational CMC presented in this chapter, and also, perhaps more importantly, to deepen and widen the understanding of intergenerational CMC between people in their early adulthood and those aged 60+.

Conclusion

In the opening paragraph of this chapter we highlighted the relevance of CMC for older people and raised two questions. First, how is the 'mobility turn' in social science (Sheller and Urry 2006; Hannam et al. 2006) addressed in CMC research with older people? Our literature review has shown that the mobility of younger generations predominates in much CMC research with older people, who tend to be conceived of as being bounded in space. Our literature review also revealed that children and grandchildren are the contact initiators of many intergenerational CMC encounters, and that new CMC tools for facilitating intergenerational CMC conversations within families have been designed. In both cases, older people seem to be portrayed as not being able to use contemporary CMC tools or using them in 'extraordinary' ways. The second question this chapter addressed is: What can CMC studies with older people offer to the understanding of mobility and ageing? Contrary to previous research, the tales have shown that our participants are not bounded in either space or time. Instead, they were the protagonists of intergenerational encounters, sometimes on the move, enabled and mediated by ICT and CMC tools that most of us already use. The tales also reinforce the need to see ageing as a lifelong process with gains and losses, and full of life experiences. Thus, this chapter reinforces the importance of CMC for older people and of developing a long-term, close-up view of how they use contemporary CMC tools. By doing so, we argue that the mobilities paradigm can enable scholars to understand further, and provide a different account of, intergenerational CMC. We plan to extend this research by exploring intergenerational scenarios of mobilities that are likely to arise out of older people's growing use of mobile apps for instant messaging such as *WhatsApp*, and emerging wearable technologies such as smart watches.

Acknowledgements

Our research would have been impossible without our participants. Thank you very much to each and every one of you. Some participants have passed away as our research progressed. We miss you and this chapter is dedicated to you. We acknowledge the support from Life 2.0 (CIP ICT PSP-2009-4-270965), Worth-Play (funded by Fundación General CSIC and la Caixa Obra Social), EEE (TIN2011-28308-C03-03), Beatriu de Pinós and Alliance of 4 Universities post-doctoral fellowships, the Ministry of Foreign Affairs and Cooperation, and the Spanish Agency for International Development Cooperation.

Notes

1 Early adulthood = 20–40 years old (Berk 2004).
2 A typical stereotype when it comes to older adults and ICTs (Giles and Gasiorek 2011).
3 This is in stark contrast to studies of mobile phones and older people. Mobile phones have become gradually incorporated into the everyday lives of most of them (Fernández-Ardèvol and Ivan 2013) and their mobile phone usage might not be too different from that of adult people (Conci et al. 2010).
4 Courses lasted between 3 and 6 months, were planned in advance and several technologies were used by the same group of participants. Workshops, however, were hands-on sessions organised to explore specific aspects in which participants showed interest as the fieldwork activities progressed.
5 The translation has sometimes kept expressions in the participants' mother tongue in an attempt to write more realistic tales. The translated version in English is shown between brackets.

6 Digital inclusion and public space

The effect of mobile phones on intergenerational awareness and connection

Dave Harley

Introduction

This chapter uses the 'mobilities' lens to explore generational differences in terms of behaviour and attitudes surrounding mobile phone use in everyday public spaces. The mobile phone is a ready accomplice to all forms of contemporary mobility, from the everyday and mundane activities within a given neighbourhood through to the global travels of the 'kinetic elite' (Graham, 2002). As a communication device it provides a source of perpetual contact with significant others that enables the ongoing maintenance of emotional connections whilst on the move, enhancing feelings of physical and virtual proximity. Urry (2007) has suggested that an underlying motive for contemporary mobilities is the deep-seated human need for physical proximity with our significant others, within what others have described as an 'ontology of connection' (Bissell, 2013). In short, our desire to be close to others drives our need to travel. Understanding the mobile phone's relationship to intergenerational mobilities provides a view into the underlying dynamics of connection which motivate travel and define the interactional mores of co-presence travelling to and within everyday public spaces. In particular, this chapter focuses on the ways that different generations negotiate and prioritise their physical versus virtual co-presences whilst travelling through public city-centre spaces, examining the effect of mobile phones on experiences of intergenerational awareness and connection in these spaces. The study uses participant observation to draw out patterns of embodied techno-social behaviour in relation to different generations' use of such technologies and then explores the underlying issues further with interviews, highlighting points of possible conflict and misunderstanding between generations. Implications are drawn in relation to the prioritising of physical versus virtual proximity, ongoing community cohesion, the design of future interactive public spaces and the need for a digitally inclusive approach to such spaces, which will accommodate all generations.

Background

The pervasiveness of the mobile phone is now hard to ignore. On the basis of mobile subscriptions, it is estimated that 96 per cent of the world's population now have access to a mobile phone (ITU, 2014). From its inception it has had a profound effect on social conduct in public spaces, with many viewing its appearance as an annoying transgression of pre-existing social norms, particularly in enclosed spaces like restaurants, cafes, libraries and airports (Wei and Leung, 1999) as well as on public transport (Monk et al., 2004), but also as a general attitude to their use in all public spaces (Ling et al. 2001). The acceptability of mobile phone conversations in the midst of ongoing communal activities still remains a contested social norm (Gant and Kiesler, 2001) and something that exists across cultures (Campbell, 2007). The increasing use of mobile phones in public spaces has led to a pervasive form of 'inattentional blindness' (Hyman et al., 2010) in which mobile phone users prioritise virtual proximity, becoming less aware of their immediate situation and ignoring immediate social signals from others, events and even physical objects (Nasar and Troyer, 2013).

Gergen (2002) has argued that what makes the public use of mobile phones so problematic for others is the division of conscious attention that they require, forcing a schism between the public norms of social conduct in the immediate vicinity and the private norms required to communicate with distant others or to interact with the mediating device. Gergen has used the term 'absent presence' to describe the way in which users of mobile phones become physically present but socially absent. The growing popularity of the Smartphone in recent years has introduced further opportunities for absent presence by integrating the functionality of previous devices and allowing mobile connections to existing Internet resources. Smartphones provide more potential for distraction by integrating the music and game playing abilities of earlier devices. When used in this way they can accentuate the need for users' privacy by cutting them off from social contact completely, creating 'telecocoons' (Habuchi, 2005). In addition, smartphones have increased the social potential of the mobile phone by extending the reach of social networking sites and social media, which had previously been bound to the home or work place. In a recent UK survey 54 per cent of the population were found to own a Smartphone, with 59 per cent of them using these to maintain their connections to social networking sites (Ofcom, 2012).

Beyond identifying the basic disruptive character of the mobile phone we understand very little about how the demands of physical versus virtual co-presence are negotiated in public spaces through the enactment of social norms, and indeed where the fault lines of social misunderstanding might appear. Given the asymmetrical nature of mobile phone uptake across the generations (Ofcom, 2012) we might expect this to be an intergenerational issue as well as a social one. In this chapter it is generational differences that are considered in terms of the social practices that surround mobile phone use. There is also an exploration of the possible repercussions that these may have for intergenerational awareness and community cohesion.

The adoption and acceptance of the social practices surrounding mobile phones is clearly something that varies from person to person. A number of factors are likely to affect this acceptance, including previous experiences with similar technologies, the expectations and demands of social context and pertinent socio-demographic variables such as age, education and socio-economic status (Rogers, 2003). In this study we explore the specific effects of age on the acceptance of mobile phone norms, such as absent presence, and consider how these play out in relation to younger and older generations.

Younger people will tend to be early adopters of new technologies and accompanying social practices (Rogers, 2003; Venkatesh, 2014). This has certainly been the case with the mobile phone and more recently the Smartphone (Pheeraphuttharangkoon et al., 2014). In the UK 97.5 per cent of young adults (16–34 year-olds) now own a mobile phone whilst only 67.5 per cent of those over 65 do.[1] When we compare Smartphone ownership the difference is even starker with about 80 per cent of young adults owning one but only 7 per cent of those over 65(Ofcom, 2012).[2] Yet these figures only provide a picture of intergenerational differences in terms of ownership of the physical devices. When we consider the attitudes that accompany mobile phone ownership and use, generational differences start to become apparent. Of the 16–24 age group, 79 per cent consider their mobile phone to be their main point of contact with others (Ofcom, 2011). Studies across Europe have shown that teenagers and young adults are highly dependent on their mobile phones and emotionally attached to them, prioritising virtual proximity when amongst strangers, keeping their phones at hand constantly and frequently using them in public spaces (Vincent, 2005). In contrast only 12 per cent of those aged 65–74 consider the mobile phone as their main route to social contact and only 5 per cent of those over 75 (Ofcom, 2011). Studies exploring older people's attitudes to mobile phones suggest that they see them as very different kinds of devices than do their younger counterparts, and tend *not* to prioritise virtual proximity except in emergencies and even then only to access immediate physical support. Their mobile phones are principally employed as devices for ensuring safety and security on specific journeys outside of the home (Hassan and Nassir, 2008; Kubik, 2009; Kurniawan, 2008) and as such are generally not used for extended conversations or even turned on when not in use (Kubik, 2009).

One would suspect that these quite distinct interpretations of purpose and proximity demands would be mirrored in terms of differing social norms between the youngest and oldest generations and their expectations of what constituted appropriate mobile phone use in public spaces. So far there is little research to verify whether these differences exist, whether they are age- or generation related and if so how they are reconciled (or not) between generations. One study by Turner et al. (2008) has shown that increasing age does correlate with negative attitudes to public use of mobile phones in places where freedom of movement is curtailed such as the work place, bars, restaurants and on public transport. However, this study was limited in that it used questionnaires to assess age differences and had a very restricted, student-based sample with ages ranging only from 17 to 43 years.

It has been argued that local communities are increasingly individualised (Lash, 2002) with their members often leading parallel lives where they do not meet one another (Cantle, 2004). One particular feature of such communities is that everyday meetings between the oldest and youngest generations within a community are now rare (Williams and Nussbaum, 2001; Vanderbeck, 2007). In this study the influence of generational stances on mobile phone use as a mediator of intergenerational contact and awareness is considered.

Methodology

In order to understand the intergenerational dynamics of mobile phone use, this research used an ethnographic approach with a view to explaining each generation's techno-social practices in relation to mobile devices 'from the inside'. Investigations employed participant observations and semi-structured interviews in public spaces in the city of Brighton in the south of England. Observations were carried out by the author in a number of different public spaces (see locations below) with the author taking field notes whilst travelling into the city centre on public transport and whilst sitting in public spaces within the city. Observations were used to identify broad generational patterns of mobile device behaviour in terms of accompanying body language, use of personal space and other behaviours relating to face-to-face interaction. Observations lasted for 21 hours overall, taking place from 19 June 2012 to 10 July 2012 and involving approximately 120–150 people. Subsequent interviews were conducted by the author in location 1, where a stall was set up in the middle of the walkway. Potential interviewees were invited to take part using a small banner with the words: 'what do you think about using mobile phones in public spaces?' Participants were self-selecting and interviews were conducted in situ. They were recorded with a digital sound recorder with permission being gained to use the interview data at a later date. Twenty-one people were interviewed, all of whom were Brighton residents. Their ages ranged from 25 to 76 but willing participants were predominantly young adults and the elderly (mostly students and retired people). This sampling bias reflected the time of day that the interviews took place (early in the afternoon on a working day). However, it did give an opportunity to compare the attitudes of older and younger generations and consider the different ways that they approach the sharing of this public space. Distinct generations were identified in line with Erikson's life stages (Erikson, 1963) with the following broad categories: childhood (under 13); adolescence (13–18); young adulthood (18–35); adulthood (35–65) and old age (65+).

The locations

Previous research has highlighted the significance of specific public locations in determining social norms for public mobile phone use (Turner et al., 2008). With this in mind four different locations (see Figure 6.1) were used as the

Figure 6.1 Clockwise from top left: Location 1, a paved walkway and sitting area adjacent to a well frequented local landmark (the Royal Pavilion); Location 2, a bus travelling through the centre of the city; Location 3, a train travelling into the centre of the city; Location 4, a communal space and sitting area outside the public library in the centre of the city

basis for initial observations. The interviews were conducted in the first of these locations as this was the busiest and provided the best opportunity to engage participants for later interviews.

Each venue provided users of mobile devices with particular attentional dilemmas derived from the physical and social nature of each context. For instance, sitting on a bus or train (locations 2 and 3) provided a physical environment that was conducive to physically interacting with the device but where the social demands of co-present others were more of an issue (for example being overheard because of physical proximity). In contrast the pedestrian areas (in locations 1 and 4) provided sufficient space for social constraints to be less of an issue but where the constant movement of others complicated physical interaction with the device. In these locations there was a constant dividing of attention between navigating the space whilst walking (or cycling) and using the interface of the mobile phone. Here the concern is with identifying generational differences in terms of their expectations of co-present interaction (physical versus virtual) as expressed through different mobile phone behaviours and expected norms of public mobile phone use in each location. From a mobilities perspective this shows how mobile phones and their inherent forms of connection come to shape the journeys (and the interactional quality of these journeys) for different generations as their intersecting trajectories converge in these particular public spaces.

Generational behaviours

The observations made it possible to identify mobile phone behaviours that were typical of each generation. Here the distinctive features of each generation are described.

Across the ages: degrees of multitasking

Members of all generations were seen using mobile phones in public places at some time during the observations but it was most prevalent amongst adolescents and young adults. A clear difference in terms of generational use was in each generation's ability to interweave their use of mobile devices with other ongoing (and off-device) activities. In locations where adolescents and young adults were walking (or cycling) through the city they would continue to do so whilst texting or talking on their mobile phones, showing a high degree of divided attention and a propensity to multitask. In contrast, older users (those in old age) in these same locations were more likely to stop everything else they were doing in order to use their mobile phones, particularly if texting. Adults (those between 35 and 65) showed varying degrees of multitasking ability when using their phones, often walking slower or sitting down when using their phones.

Adolescents (13–18): the seamless social network

Gatherings of teenagers in public places were often accompanied by simultaneous use of mobile phones, iPods and/or portable gaming consoles. This was observed with college students (aged 15–17) travelling on the train (in location 3) and in the centre of the city (in location 4). In these gatherings they would sit or stand in a circle with each individual seemingly focussing their attention on their individual screens. Whilst this might appear to be an isolating activity in terms of removing direct eye contact from one another, in other ways the sense of a social meeting was maintained. Whilst their visual attention may have been occupied their auditory attention remained available to one another with conversations continuing despite ongoing and simultaneous texting or updating of social networks. At times they would share their online activities with one another creating conversations that incorporated (absent) others via their recent text and photo postings to social networking sites. This resulted in what might be called a 'seamless social network' in which online social activities were integrated into immediate person-to-person (rather than face-to-face) gatherings and vice versa.

Young adults (18–35): phone as companion and the public/private bubble

Groups of young adults in public did not engage in the seamless social networks of their younger counterparts but tended to exclude those present from their conversations by removing eye contact and orienting their bodies away from any congregation whilst talking on their mobile phones. Mobile phone conversations were not curtailed because of this exclusivity but were accepted by the rest of the group without complaint, creating a public/private bubble. For those receiving calls there was variation in the prioritisation of mobile phone voice calls over immediate face-to-face interactions. Different levels of discretion were used and this would dictate how much they lowered their voice, avoided eye contact and/or withdrew from the centre of the congregation whilst talking on their phone. Text messaging was less common amongst this group when they were out with others in a public space. Young adults were the most visible users of mobile phones in public places. A distinctive aspect of their presence in these places was that they were often seen alone. In these instances, the mobile phone was carried in hand and displayed as a symbol of social status implying continual social availability and connection, i.e. they were not really 'alone'. In such a way the mobile phone appeared to act as a constant companion for many solitary young adults.

When alone it was common for this age group to talk for extended periods of time whilst engaged in simultaneous activities that coincidentally involved other people who were present in the local vicinity. For instance, in location 1 it was not uncommon to see young adults walk the entire length of the walkway (about 200 metres) talking or texting on their mobile phone. This meant looking up occasionally from the phone but otherwise ignoring all passers-by and opportunities for immediate social contact. Those prioritising mobile phone contact over all else would carry out extended conversations at high volume with no acknowledgement of the public context around them.

Adults (35–65): efficiency in between moments

Mature adults were observed using their mobile phones at moments in-between activities or places. Mobile phone activity (whether texting or talking) occurred at particular thresholds just prior to entering buildings such as upon leaving the library or entering a convenience store (both in location 4). This group was more likely than other age groups to be seen talking on their phone when alone. Phone use occurred during times that would have otherwise been taken up with 'mindless' activity such as walking and which made efficient use of available thinking time.

Old aged (65 +): discreet use

For the most part older people's use of mobile phones was either non-existent or hidden from public view. Out of all the older people observed, only three were seen using their mobile phones in public. On each of these occasions their use of the technology was what one would call 'discreet'. They would stop whatever else they were doing, take their phone out of a bag or pocket, use it for a distinct purpose and then return it. This appeared to be driven by a need to attend to a single task at a time but was compounded by difficulty in seeing the mobile phone when moving. Some older adults would put on glasses or adjust them in order to read and operate the phone. Those phone conversations that were observed were short and to the point.

Generational attitudes

Whilst observations provided a view of different generation's behaviours in relation to mobile phones, interviews exposed the underlying attitudes that underpinned these behaviours and coloured interpretations of others' behaviour.

All of those interviewed owned a mobile phone irrespective of age, but there were clear differences in terms of justifying ownership and use. For younger adults, there was an assumed necessity to own a mobile phone. It provided a conduit to social realities that were unquestionably important in sustaining social identity and personal equilibrium in everyday life, wherever they were. The phone itself had to be turned on and close at hand either in a pocket or more likely in the hand. They had to be socially available through the phone at all times. As one interviewee explained the significance of his phone: 'I think it's to be in contact, just that simple, I think that's about it' (p. 6).

Perpetual contact (Katz and Aakhus, 2002) provided more than just the possibility of social contact, however. As observations suggested, the mobile phone acted as a 'constant companion', assuaging feelings of social anxiety when strangers were present (i.e. in public spaces): 'It's a little comfort blanket … a comfort brick of technology. It makes you feel safe, your phone' (p. 9).

The phone was used as a mental companion when interaction with friends was not possible or the individual was alone in a public place where strangers were present, such as waiting at a bus stop or sitting on a train. Activities engaged in included 'non-interactive' social activities like perusing Facebook posts and text messages as well as more individual activities like playing games, creating lists and using the phone as a notebook.

In line with previous studies these younger adults were highly dependent on their mobile phones (Vincent, 2005), experiencing frustration and panic when unable to connect to 'the network' (either mobile phone network or the Internet) for some reason: 'the panic you get if you leave your phone somewhere and you don't have it on you. I keep it in my hand so I get all my notifications' (p. 9).

In contrast, older mobile phone users were not keen to be constantly connected and preferred to keep their mobile phones turned off unless they were expecting or making a call. Their comments fitted well with observations of 'discreet use'. In line with earlier research (e.g. Kubik, 2009) they justified owning a mobile phone in terms of it being an emergency resource that they could use should the need arise. As one interviewee put it: 'It's in my bag now but coz I always take it with me coz you never know when you might need it but really mobiles are for emergencies' (p. 1).

However, these older people's use of mobile phones had developed beyond emergency use. One major use for mobile phones was as a logistical tool that enabled meetings with friends and family:

> Why I have one [a mobile phone] is because … sometimes, I meet a friend in Worthing – he drives and there may be a problem so if we both have our mobile phones with us if there's a difficulty we can text each other … sorry I'm held up in traffic or whatever.
>
> (p. 2)

Despite developing some of their own uses for mobile phones there were still obstacles to more extensive use. These included age-related constraints on use such as difficulty reading the screen and hearing incoming calls (particularly in crowded places); an unwillingness to fund greater mobile phone use, a general distrust in technology, and an overriding preference for direct face-to-face interaction given the choice.

Interestingly, the younger adults expressed a preference for interacting via text-based forms of communication, one saying that 'texting is cheap, well its free, it's easy communication' (p. 3) and another that it was about 'maintaining a connection in an effortless manner' (p. 6). Some found that face-to-face interaction with people they did not know whilst waiting for a bus or in a shop was an odd thing to do and actually made them feel quite uncomfortable.

Differences over social norms

There was a lack of observable disagreement between these generations when it came to their public use of mobile phones. This seemed to suggest some tolerance over one another's behaviour. However, the interviews with our older participants (and one of our younger interviewees) revealed that they were ill at ease with certain aspects of mobile phone use and (at the same time) felt largely powerless to express any opinion about this. Older people were particularly troubled by the now common practice of younger people conducting their private and trivial conversations in public spaces. As one interviewee said in relation to overhearing other people's phone conversations: 'So everybody knows they're going out to Al Forno's [a local restaurant] tonight. I mean I wouldn't like everybody to know my business' (p. 1).

The discomfort increased when they were stuck in confined public spaces with a mobile phone user. Here there was no way of escaping the act of listening but they felt that there should be a greater respect for one another's personal privacy. Examples given of such locations included trains, buses and libraries: 'I do think mainly when you're on the bus or the train … the library, it's not like the old library – silence please – you get human voices and other people shouting on the phone. I do think it's annoying' (p. 2); and even public toilets: 'I've noticed that you'll go in the toilets somewhere and someone will be speaking and he's on the mobile phone. I think – you're not concentrating on what you're doing yourself and it's an insult to the person you're speaking to' (p. 2).

Most of the young people interviewed did not see other people's use of mobile phones in public spaces as problematic, nor did they see it as a generational issue. As one interviewee explained, 'I think a lot of people are very civil though I mean when they talk on the phone they're very, they are quiet. I mean personally it doesn't bother me coz they don't talk in an obnoxious manner so it's really OK' (p. 6).

Similarly, their own pervasive use of the phone was not seen as problematic. Indeed, it was even seen as an achievement by some of our interviewees: 'I walk down the street Facebooking on the phone, I'll be reading my emails, texting, calling, facetiming, everything … So yeah, nothing really stops me … whilst speaking to people, Facebooking, whatever, yeah. I use it in the toilet. Basically my phone goes everywhere with me – in the shower' (p. 9).

There was only one younger person (albeit at the older end of the spectrum) who acknowledged that his own and other people's phone conversations might impinge upon other people's need for privacy and he was keen to point out that those younger than him did not behave in the same way:

> If somebody calls me up on a train that I haven't spoken to for a long time, I'd say have a very brief chat with them and would want to continue that chat with them at home or somewhere more private. I think as people get younger and younger below me that sense of privacy and disclosure and time and place is so much less of an issue to them, they're quite oblivious to it.
>
> (p. 5)

On the face of it most of the younger people who were interviewed did seem to be happily departing from the expected social etiquette in public spaces. However, there were times when even their own behaviour was deemed to have gone too far in terms of ignoring the immediate social expectations of their peers: 'About five of us all had our phones out and we were all scrolling through stuff. We've even been sat in the same room and been talking on WhatsApp! Whilst we were all in the same room' (p. 9).

In certain situations, the extreme nature of these mutual absent presences became untenable and they were forced to interact directly with one another. This was most often the case when taking part in a shared activity like going to the cinema, going out for dinner at a restaurant and nightclubbing.

Discussion

Clearly the social norms around the public use of mobile phones are changing, and indeed have already changed as a result of the pervasiveness of the mobile phone and the arrival of the Internet on these devices. This study shows that there are important differences between younger and older generations when it comes to their expectations for public use of mobile phones and the attitudes that underpin their everyday use of these technologies. These differences can act to negate generational awareness and interaction, with them effectively being absent to one another even though they are travelling through the same everyday public spaces. There is a generational shift towards 'the network' and prioritising virtual co-presence as the source of social reality, and this has serious implications for local community cohesion within neighbourhoods such as those studied here.

Mobile phones are central to younger people's sense of involvement in life, and during this study they were constantly on show and in use, with continuing use in almost all public spaces. The phone served as a social entity in itself, providing companionship even when not being used for social interaction. For the older people in this study the phone remained a discreet tool with much less significance in maintaining social involvement. Observing all generations together suggests that whilst there are differences in terms of how each generation appropriates the mobile phone there is an increasing dependence on mobile connectivity for younger generations to sustain their social lives. Perhaps more importantly, this study has shown that the meaning of what constitutes shared everyday public space is shifting from one (expected by the older generation) which is based on immediate physical co-presence as a source of social interaction and community involvement, to one (expected by younger generations) which prioritises 'the network' and virtual co-presence as the primary source of interaction and affiliation. Whilst there does appear to be a great degree of tolerance when it comes to accepting these different generational perspectives, it is clear that different generations now inhabit distinct social spaces even when they are travelling through the same physical locations or neighbourhoods.

Generational attitudes to the meaning of shared public space have clearly diverged, and by inhabiting different social realities they remain invisible to one another. As previous research has shown (Wei and Leung, 1999), this divergence is experienced most extremely in confined public spaces such as on public transport and within public facilities such as toilets and libraries. Within these spaces there are real difficulties in resolving these different perspectives and of providing a bridge between disparate generations. So far our ability to resolve these differences in attitude and behaviour are poorly developed. For some older people it appears as if younger people have lost their sense of correct behaviour entirely and that they themselves are becoming further isolated from modern life. At the same time, it seems that younger people are becoming fearful of face-to-face contact with strangers and would rather retreat into the relative safety and 'ease' of their mobile/Internet connections. Interestingly, the interviews also highlight that younger mobile phone users do continue to have rules of social decorum but these only surface when they are face-to-face with their mobile (virtual) contacts. This suggests that younger people are indeed operating through an 'ontology of connection' (Bissell, 2013) where the drive for physical proximity with mobile (virtual) contacts prevails. Meanwhile older people appear to be operating through what has been called an 'ontology of exposure' (Bissell, 2013) in which they remain open and receptive to their 'near-dwellers' whilst travelling through their neighbourhood. In mobility terms it is therefore interesting to note that whilst different generations are making similar journeys through everyday public space, even engaging in similar daily routines, they are making themselves available to quite distinct social realities – one physical and the immediate and the other virtual and mobile with a promise of future physical closeness.

Public spaces within cities are set to become increasingly digitised as part of moves to create 'Smart Cities' within Europe (Caragliu et al., 2011). Here public space will be further augmented by mobile/wireless connectivity (UK Government, 2013) and interactive technologies. It is likely that public spaces of the future will enlist mobile phones as conduits for establishing an awareness of city dynamics and for user-based interaction and community involvement (e.g. Ballagas et al., 2006). Understanding these generational differences is important if we are to engage all members of a given community equally in such future cities. Those spaces that harness the ubiquity of mobile phones for public interactivity will have to accommodate these different uses and expectations in relation to mobile phone use if they are to be truly inclusive spaces.

Conclusion

Through participant observation and interviews this study has shown how attitudes and behaviours surrounding mobile phone use in public space are diverging across generational lines. The journeys of younger people take place with a backdrop of perpetual virtual connection and co-presence, whilst older people's journeys remain open to the possibility of immediate physical co-presence interaction. Whilst there is a general tolerance of public mobile use, there are points of misunderstanding and potential conflict between generations, particularly in confined public spaces where there is less freedom of movement or expression. In such situations it is difficult to express or share a contrary opinion and it seems that we have yet to develop any social or technological means of addressing this form of social disturbance in a reasonable and equitable manner. These generational dynamics have implications for ongoing local community cohesion, and we should be exploring ways of connecting generations and providing opportunities for each generation to voice these differences. Public notices are sometimes used on public transport in the UK to limit mobile phone use. However, the older people in this study suggested that these were ineffective. Given that younger generations are now living their lives through a virtual social world, this is perhaps unsurprising – they are no longer paying attention to the social signals of the physical world so this will not work. There needs to be a means of developing a social awareness that can bridge the physical and virtual worlds and which will reconnect every-day travels with the serendipity of meeting others (of all ages). This could be done technologically through either context-aware mobile phones or through mobile phone-aware public spaces, i.e. where quiet, mobile-free spaces were embedded into the fabric of public buildings and public transport. Alter-natively, social etiquette around mobile phone use may change over time and allow a more open dialogue to take place about where and when their use is appropriate. This would encourage greater understanding between different generations but also, it seems, between members of the same generation. Such considerations should be central to the future design of interactive public spaces arising through Smart City initiatives, if we are to maintain a digitally inclusive approach to such spaces that will accommodate all generations.

Notes

1 This can be broken down further with 80 per cent of 65–74 year olds and 55 per cent of over 75s owning a mobile phone.
2 This can be broken down further with 12 per cent of 65–74 year olds and only 2 per cent of over 75s owning a Smartphone.

7 Talking about my generation

Emigration and a 'sense of generation' among highly skilled young Italians in Paris

Hadrien Dubucs, Thomas Pfirsch and Camille Schmoll

Introduction

Along with other countries in southern Europe with histories of emigration, Italy has recently experienced a new wave of emigration, primarily of young, skilled professionals with mobilities quite unlike those of previous generations of Italian migrants. This new southern European emigration has generated intense discussion in public debate and the media, but it remains little explored in academic studies, until in very recent times. We look at this new emigration through the example of young, highly skilled Italians in Paris, a city of particular interest as a leading destination of both Italian mass emigration in the period 1860–1970 and new, skilled mobilities in recent years.

Economists have conducted most of the few studies analysing this new Italian emigration, taking the 'brain drain' or 'brain circulation' approach and focusing on labour market factors. We argue for a broader approach in this chapter, especially through demonstrating that young Italian adults' emigration is not simply due to labour market factors, but also to a 'generation gap' in Italian society (Del Boca and Rosina, 2009) that makes it difficult for youth to access independence and social recognition. A strong 'sense of generation' (a feeling of belonging to the same generation) emerged spontaneously from the interviews we conducted in Paris. In this chapter we highlight that this 'sense of generation' is both a factor in migration and a consequence of it. The sense of belonging to a 'sacrificed generation', confronted with a dearth of recognition in a society with a poor meritocracy, not only prompted skilled young people to emigrate but went on to be reinforced by the migratory experience. Although this generational perspective is quite present in the media and southern European migrant cyberspace (through what are now well-known expressions such as 'the 592 euro generation' in Greece or the 'nobody generation' in Italy; Chucchiarato, 2011), it has been little explored in migration studies, which have examined intergenerational relations[1] but not a 'sense of generation' connected to migratory experiences.

Our choice of the term 'generation' draws on the sociological work of Karl Mannheim (1952 [1923]), which defines a generation as 'a group of individuals whose members have experienced a noteworthy historical event within a set period of time', even if they are not exactly the same age. In this respect, young Italian migrants can be seen as part of a 'crisis generation' that emerged after several structural changes in Italian society, as we shall see below. It is also a European generation, where young people's trajectories include the experience of the possibilities and limits of EU integration and freedom of movement. Similarly, the concept of 'youth' is not solely based on age. Following Mauger (2010) and other sociologists, we consider youth as 'the phase of the biographical trajectory in which individuals have not reached yet a stable position in both labour and matrimonial markets'. It is the time it takes to find one's place in society. Youth is defined as much through social position and family status as it is by age. This is particularly relevant for southern Europe, where transition to adulthood is long and difficult. In Italy this is often called 'delay disease' (Sgritta, 2002): delay of departure from one's parents' home, marriage age, the birth of the first child (which often takes place after the age of 35), and so on.

In keeping with this broad definition of youth, the migrants we studied in Paris were aged 20 to 40. Our methodology is mainly qualitative. In 2012–2013, we conducted twenty in-depth interviews with young, highly skilled Italians (university graduates), aged between 25 and 40, who had been living and working in Paris for at least one year. We explored their biographical and family backgrounds and their varied motivations and identities in addition to their migratory and professional trajectories. Although this new emigration is difficult to measure and often invisible to official statistics (since it takes place within the Schengen space), we were also able to use the limited quantitative data available in both Italy and France: the French national census (INSEE, 1999 and 2009) and the Italian population registers (especially the AIRE: 'Anagrafe degli Italiani residenti all'estero'). Last, as the increasing use of mobile media has become crucial to the contemporary experience of mobilities, we analysed representations of the new Italian emigration in the media and in migrants' rapidly growing cyberspace. The migrants' 'sense of generation' is largely maintained and nurtured through virtual communities, blogs and websites that have been proliferating for many years in conjunction with the new Italian emigration. The 'sense of generation' is closely connected to four main issues for these young Italian migrants, each of which will be examined in turn: a new phase in Italian migration history in France, European integration (a generation of 'European movers'), the labour market crisis (a 'jobless generation'), and the generational divide within Italian society (the '*generazione nessuno*', a 'nobody generation' lacking recognition and suffering from marginalisation in Italian society).

'Old' versus 'new' Italian migrants?

Italian migration to France has a long history, well predating the beginning of mass migration in the 1860s (Corti, 2003; Miranda, 2008; Sanfilippo, 2011). Recent historiography has shown that an initial migratory system between the two countries appeared during the Renaissance. It was based on temporary professional movements within a few limited economic sectors by both the skilled (artists and craftsmen circulating between cities and royal courts) and unskilled (northern Italian farmers migrating to the adjacent southern French countryside for the harvest). These old flows continued uninterrupted into the contemporary period, but the former was overshadowed by the amplitude of unskilled mass emigration from Italy that emerged gradually in the nineteenth century. This second migratory system was mainly composed of unskilled farmers from both northern and southern Italy, moving not only to rural southern France but to its large northern and eastern industrial cities (Paris; mining areas such as the Metz and Douai regions) as well. France was the main European destination of this nineteenth- and twentieth century Italian mass emigration. Between 1876 (the date of Italy's first official census of emigrants) and 1976 (the year in which returns outnumbered departures for the first time), France took in 4.3 million Italian citizens, almost one seventh of the 29 million Italians who left their country in this period (Rosoli 1976). Since the end of the 1970s, however, a new phase emerged. According to official data, Italian entries into France have decreased substantially in the last three decades, as did the number of Italian citizens living in France: although the total presence fell from 252,000 in 1990 to 173,000 in 2009 (see Table 7.1), the decline is in fact primarily due to the naturalization of older generations of Italian migrants. The Italian population registers (AIRE), which allow annual flows to be measured, actually show that Italian migration to France has remained roughly stable for the last 20 years, with 3,000–4,000 entries per year until the end of the first decade of the 2000s. It then increased significantly under the euro crisis of the period (which hit Italy hard), with 5,000 official registrations in 2011 (see Figure 7.1). We can thus see that although mass emigration is clearly over, strong and steady Italian migration to France continues, albeit very fluid, difficult to measure, and under-researched.

Yet the young Italians we interviewed in Paris do not spontaneously evoke this long history, and do not seem to be particularly aware of it. They do not consider themselves heirs of a tradition of Italian emigration, nor are they connected to the social and regional networks of older generations of Parisian Italians, which they describe as 'another world' (Isabella, 40, civil servant). This social and cultural distance reflects genuine and deep differences between

Table 7.1 Italian citizens in France: 1990–2009

	1990			1999			2009		
	France	*Paris region (Ile de France)*		*France*	*Paris region (Ile de France)*		*France*	*Paris region (Ile de France)*	
Italian citizens	252,000	43,000	17%	200,000	36,000	18%	173,000	43,000	25%

Source: INSEE, National censuses of 1990, 1999 and 2009.

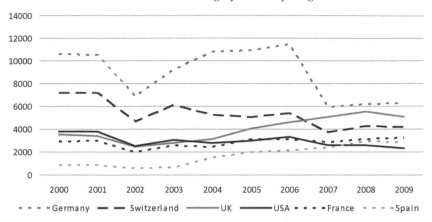

Figure 7.1 Annual flows of Italian migrants to main foreign destinations (2000–2009)
Source: Fondazione migrantes, Rapporto Italiani nel mondo 2012, p. 34 and 2013, p. 41 ('trasferimenti di residenza per l'estero')

Italians who moved to France recently and those of the previous waves of migration.

In comparison with the older migrants who were often from rural communities, new Italian migrants to France are mostly urbanites attracted to Paris and large metropolises in France and Europe-wide (see also Gjergji, 2015). A quarter of them are now living in the Paris metropolitan area, compared to only 17 per cent in 1990 (see Table 7.1). The same metropolization of migration has been observed in Germany, where the share of total of Italian migrants in Berlin has increased steadily in recent years (Del Prà, 2011). This attraction to large cities is closely linked to the socio-professional profiles of migrants. It is also essential to emphasize that they are more often than not university graduates and hold highly skilled jobs. Figure 7.2 shows the increase in the number of professionals and highly skilled workers among Italian migrants in Paris over the last ten years, while the number of manual workers was halved over the same period (see also Table 7.3).

Most respondents moved to France in the final years of their university studies, or just after completing them. This is another difference with older generations of Italian migrants, who came to France very young (usually under the age of 20). This seems to be a general trend: in recent years the proportion of people under 19 in the official count of Italian migrants in France has tended to decrease, falling from 26.6 per cent in 1981 to 17 per cent in 2009 (INSEE). Another distinctive feature of our respondents is that they are single female and male migrants who moved to France alone (see also Gjergji, 2015). This is illustrative of a broader dynamic of the individualization of migratory trajectories, where migrants are increasingly independent of regional or family-based channels of migration. Before settling in Paris, however, most of them had previously spent several months in France as students, using university networks and European exchange programmes.

Table 7.2 Annual flows of Italian migrants to main foreign destinations (2000–2011)

	2000	2001	2002	2003	2004	2005	2006	2007	2008	2009	2010	Total
Germany	10,620	10,518	6,848	9,191	10,768	10,927	11,464	5,939	6,185	6,281	4,803	**93,544**
Switzerland	7,188	7,217	4,672	6,161	5,236	5,042	5,407	3,710	4,262	4,196	4,619	**57,710**
UK	3,501	3,422	2,439	2,795	3,123	4,062	4,624	5,087	5,528	5,042	5,251	**44,874**
USA	3,772	3,793	2,557	3,072	2,797	3,003	3,356	2,574	2,591	2,345	2,557	**32,417**
France	2,918	2,980	2,032	2,601	2,487	3,130	3,134	2,881	3,135	3,248	3,784	**32,330**
Spain	854	893	622	685	1,541	1,985	2,155	2,398	2,924	2,890	3,036	**19,983**

Source: Fondazione migrantes, Rapporto Italiani nel mondo 2012, p. 34 and 2013, p. 41 ('trasferimenti di residenza per l'estero')

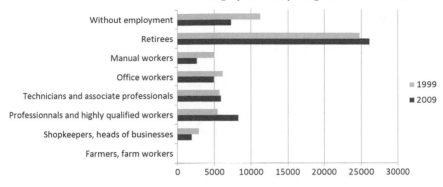

Figure 7.2 Socio-professional categories (1999–2009) amongst Italian migrants in France
Source: INSEE, French National Census.

Table 7.3 Socio-professional categories (1999–2009) amongst Italian migrants in France

	2009	1999
Farmers, farm workers	11	8
Shopkeepers, heads of businesses	1,912	2,917
Professionnals and highly qualified workers	8,251	5,437
Technicians and associate professionals	5,837	5,680
Office workers	4,864	6,149
Manual workers	2,632	5,024
Retirees	26,045	24,771
Without employment	7,251	11,262

Source: INSEE, French National Census.

An 'Erasmus generation?' Paris as part of broader European circuits

Migrants' 'sense of generation' is connected to the growing mobility of youth within the European Union. Our respondents describe themselves as a generation of 'European movers' who grew up under EU integration, with mobilities, identities and trajectories very different from those of older generations of Italian migrants in France.

Consequently, migration is still not an entirely individual or self-made process, regardless of the fact that family and regional networks no longer guide international moves. Many interviewees have lived abroad on exchange programmes while in high school or university, and a majority participated in the Erasmus programme or other European mobility programmes (such as 'TIME' – Top Industrial Manager for Europe). Such student stays proved to be crucial experiences for further mobility: while abroad students learned foreign languages, gained experience in dealing with an unfamiliar environment, sometimes met a foreign romantic partner and started planning a family, and made new friendships, building international social networks. This process is well illustrated by the case of Alessio, a 39-year-old Italian who works as an economist. He spent one Erasmus year in Belgium, got a Ph.D. in the United States, and worked seven years in Montreal, Canada, before moving to Paris. He is currently based in Paris but frequently returns to Montreal for work and considers Paris as just a step in his career. Moving abroad is portrayed as something 'almost natural' in his interview.

> In La Bocconi University [where Alessio studied in Milan] everybody takes part in the Erasmus program, it is quite natural. When I was 22 I spent one Erasmus year in Leuven University, in Belgium. I loved the experience! When you start going abroad it becomes natural. Moving abroad becomes easy, almost natural. I am not sure I'll stay a long time in Paris.
>
> (Alessio, 39, economist)

The experience of European circulation has an impact on social networks, as well as on self-identification and one's sense of belonging. This is obvious in the case of Antonio, a young Italian engineer living in Paris with his French wife and their two children. He first came to France as a student on a European programme, then met his future wife at a European youth camp in Portugal. He describes himself as a 'pure product of Europe' and thinks that he belongs to a 'pioneer generation' amongst Europeans. And yet most of our respondents keep strong ties with their country of origin, and none describes themselves as a 'citizen of the world' without national or regional affiliation. They all return regularly to Italy, at least two or three times a year, especially during holiday periods (Christmas and summer), usually to their parents' home. Geography plays a role in such frequent return visits: France and Italy are neighbouring countries, well connected by low-cost airlines. Even in the cases of infrequent return visits, the connection with Italy and the family is retained through the use of email, phone calls, and Skype or WhatsApp to stay in touch. As shown by recent research, contemporary mobilities and technological connections are inextricably linked (Ponzalesi and Leurs, 2014; Diminescu et al., 2014). Mobile media are a crucial aspect of the ability of being mobile and the experience of being far from one's relatives. Such a use of communication technologies has been extensively described by research on migration in different contexts. The status of Italy thus seems to be quite ambiguous: it is unanimously considered to be an impossible place to work and thus to live, but it remains a crucial reference in defining oneself, and ties with Italy or the home region are always very strong and emotional.

> Looking at Italy is like looking at a man you really loved but with whom it is no longer working out.
>
> (Claudia, 39, academic)

As soon as I arrive in Italy I become very emotional and I want to stay there.

(Alessandra, 31, NGO manager)

Everytime I went back to Rome I had tears in my eyes. I was deeply moved by the beauty and the light.

(Isabella, 40, civil servant)

Young Italians in Paris can thus be seen as a new generation of European movers who were born into the EU integration process and use the Schengen space's new provisions for communication with ease (academic exchange programmes, low-cost airlines, etc.). For many of them, Paris is seen as a step towards a wider European or global circulation, although interviews show no contradiction between being increasingly mobile and keeping a sense of regional or national identity. Young Italians in Paris are flexible and 'plural individuals' (Lahire, 2011) with multiple identities that they easily synthesize and feel free to use according to the social context, selecting from among their European experiences Italian culture, regional roots, or ties with France and Paris (Recchi, 2014).

Sense of generation and the labour market: escaping unemployment or seeking recognition?

In addition to the European integration process, malfunctions in the Italian labour market play a key role in producing a sense of generation among new Italian migrants. In interviews respondents describe themselves as part of a 'jobless generation', without opportunities to land official contracts and top professional positions. With this self-characterization they echo the extensive coverage of the issue of 'brain drain' in the Italian media and public debate. As an example, in September 2012 the weekly news magazine *L'Espresso* devoted a special issue to *I nuovi emigranti* (the new emigrants) with the following headline: 'Young, graduates, and unemployed, they decided to take the leap and find employment outside of Italy. Just like what happened a century ago. *L'Espresso* brings you their stories, full of hope and despair.'

By describing the new national emigration as a mass exodus and focusing on the brain drain perspective, Italian media tend to homogenize young migrants abroad as a unified community, and contribute to shaping a strong sense of belonging among them. The sense of belonging to a 'jobless generation' has a particular impact on young, skilled adults. Recent studies have demonstrated that Italy is the only EU country where, among young adults, the probability of being unemployed increases proportionately to the education level (Ballatore, 2007). This is partly due to the specificities of the Italian labour market on the demand side, but also to the slowing of Italian economic growth, which has not allowed educated young people to be absorbed by the labour market as higher education becomes so widespread. Following a general trend in Italy, many of our respondents anticipated difficulties in finding a job early in their professional and life plans by prolonging their education in Italy, and moved abroad in the last years of their university training without even trying to get a job in their home country (for a similar analysis applied more broadly to southern European emigration countries see Triandafyllidou and Gropas, 2014).

This is why the commonly held view in the field of migration studies that there is a direct relationship between unemployment and emigration, should be refined and discussed. The nexus between upward social mobility and spatial movement is contextually bound. Two points must be stressed. First, these migrants have rarely experienced unemployment in Italy – their very pessimistic view of the Italian labour market is mostly based on examples from friends and fellow students. The other point is that the key problem seems to be a lack of recognition, rather than a lack of work. In other words, it is possible to work in Italy, but as distinctly over-qualified and under-paid. Young graduates thus tend to consider emigration as a viable option and seek employment in countries where their skills would be better recognized. Respondents reject an Italian society that they see as failing to involve young people, to recognize their abilities and worth, or offer them career advancement. In doing so they bring back the issue of social justice at the core of their mobility decisions. They rarely single out wage differences to explain their migration, preferring to justify their decision in terms of a quest for professional and social recognition, from a strongly meritocratic perspective. Such assumptions about the 'fairness' and 'meritocracy' of the French labour market should, however, be understood in light of the difficulties they think they would have faced in Italy, rather than the specificities of the French labour market *per se*. Italy's lack of meritocracy is a recurring issue in interviews, as it appears in the following quotes (see Triandafyllidou, 2015 and also Scotto, 2015 for similar conclusions on the case of Italians in London).

> In Italy, I never got anything just because I was good. In France I did. Right away. I got a job interview at the university. There were eighty applicants and only four of them got an interview. It made me realize that France was a fair country, where you can get something even if you have no connections or 'raccomandazione' […] I worked at the Italian University for seven years. For free. I spent entire days administering examinations … I never received one euro. There was no chance for me to succeed in Italy. Not because of me, but because the system is enclosed.
>
> (Claudia, 39, academic)

> We [young skilled Italians in Paris] are all aware that if we are here it's because we manage to do things here that we cannot do in Italy, simply because here we have been put in charge at a level that is unthinkable in Italy. Really unthinkable. I think I never could have done editorial work in Italy at the age of 30, I mean 27, 28, 29. Impossible. I would never

have been a project manager for a big NGO like WWF. This is simply because the work culture in Italy does not allow it. As a young person you are not given any chance to get a decision-making position, they don't take any risks. They give you no recognition, while the French do...

(Alessandra, 31, NGO manager)

Paris is a great place. The system here is based on meritocracy, unlike in Italy. Skills and talents are recognized [...]. In Paris if you have a good resumé you can find job opportunities. In Italy it's impossible. You must have good family connections.

(Adriano, 38, artist)

Youth emigration and the 'generation gap': an Italian/southern European problem?

European integration and dysfunctional labour markets are not the only factors contributing to a 'sense of generation' among young Italian migrants. Such a feeling of injustice is also connected to the position of young people in southern European societies, which not only offer young people few opportunities for employment but also limit their access to housing autonomy, political involvement, and social recognition. In other words, far from being strictly economic, the crisis facing young Italians is multidimensional and concerns different aspects of society.

There is a feeling I share with many Italians in France. It is a deep worry about the future of Italy. I realize that we talk about it very often. Every day I read *La Repubblica* online. All of us keep looking at Italy's sad destiny. Because it is an integral part of our lives [...]. Every time we talk about Italy it is so painful for us. The current political situation is catastrophic; I really do not know where it will end up. [...] These things, like corruption, happen over and over again, it makes me sick. [...] Facebook is the most convenient way to share a sad song about Italy or to comment on Berlusconi's latest blunder. [...] 90 per cent of my friends who are left-leaning voters like me voted for Grillo.[2] All my friends in Paris aged 33 like me voted for Grillo. It is a question of generation, the only thing we can do is to wait until the elderly die. Young people are not given a voice yet [...]

(Giorgio, 33, employed in the voluntary sector)

The above interview demonstrates a frequently recurring practice: use of 'we' when talking about Italian migrants' experience, conveying the idea that one's own migration reflects a more general context for young Italians abroad. Interviewees tend to spontaneously describe their individual trajectories as being part of a collective experience. They unanimously stress the great difficulty young graduates have in starting an independent life in Italy, and in so doing make a sort of 'sense of generation' palpable.

This sense of generational belonging echoes the more generic 'generation gap' that exists in Italian society and southern Europe in general. The migratory experience enhances this sense of belonging, all these young migrants to some extent sharing the feeling of having been rejected by both the Italian labour market and society. The intense use of blogs, social networks and online forums as a public space for discussing and sharing experiences plays a significant role in nurturing and enhancing such a sense of belonging. Though they may also refer to forms of local belonging, blogs, forums, and virtual groups contribute to building the image of a transnational community of Italian migrants living abroad, sharing a common interest in Italian society but also a sense of somehow being rejected by it.

> We were young and they told us : Go to university or you'll be nothing – We did so. [...] And then they told us: Don't you know grades are useless? You should learn a trade! – So we did. [...] We did not have children. And then they told us, from their professional positions easily found in the 1960s: You are big babies, you do not want to grow up and start a family. And meanwhile we were paying for their pensions and saying goodbye to ours [...] At this point, we could not kill them, could we? So we emigrated ...
>
> (posted on Facebook by Ornitorinko, Italian migrant and blogger, 21 May 2014)

Respondents narrate with very strong words the sense of belonging to a generation facing many more difficulties than the former generation. The metaphor of 'age struggle' is repeated persistently in the interviews. It is based on a strong opposition to the respondents' parents' generation, which worked and had families during Italy's 'economic miracle' period[3] in the 1960s, 1970s and 1980s.

> In Italy, when you are 40 years old people start to think that maybe, yes, you can start to grow up, but you are faced with a very, very long period of infantilization, and this is also because people there work till they are very, very old. Newly arrived young people have to wait in line, and you pile up to the age of 40. They have always told us: you will work when you are 35 and after a 5-year unpaid internship experience ...
>
> (Alessandra, 31, executive of an NGO in Paris)

The notions of 'supernumeraries' (*surnuméraires*) or 'disaffiliated' (*désaffiliés*) proposed by sociologist Robert Castel may be useful in understanding these migrants' pessimism and sense of belonging (Castel, 2003; see also Péraldi, 2005). The 'age struggle' issue is also quite present in the media and plays a key role in political life, as seen during the last general election campaign in 2013. The generational issue in Italy was a central theme of Beppe Grillo's '5 Stelle' party, and the party's strong electoral showing allowed many young candidates to enter parliament. Similarly, Prime Minister Matteo Renzi promoted many young ministers in his government in an effort to change the image of Italian politics. However, gerontocracy is far from disappearing from Italian society and its political sphere. Migration is thus often described as a way to escape a society where intergenerational upward social mobility is no longer guaranteed. Young migrants say they face a lack of opportunities in the labour market, the housing system, and political and public life. This feeling of belonging to a 'sacrificed generation' is not merely self-representation; it is also related to current social processes. This was clearly demonstrated in recent studies such as that of Rosina and Del Boca (2011, 33): 'official data shows that Italy is one of the western countries with the oldest political class (in terms of age), and with very little generational renewal among the ruling class'. This difficulty in guaranteeing generational renewal is also quite present in the family-formation process, the so-called Italian *famiglia lunga* (Scabini and Donati, 1988; Pfirsch, 2011) – that is, young adults continuing to live at their parents' homes long into adulthood. Rosina and Del Boca propose the '3-G' theory for rethinking current social inequalities in Italy, positing that the traditional class division is also compounded by three more great divides: the geographical divide between Northern and Southern regions, the gender divide, and the generational divide. Of course, such a 'sense of generation' is not unique to Italian migrants, and it may also be strong among other skilled young people of southern Europe; this merits further discussion and research.

Conclusion

Recent Italian migration to France differs sharply from the mass migration of the period 1860–1970 in many respects. Recent Italian migrants are far fewer and more qualified, and generally move from a big city to Paris or other large cities. Above all, they differ from earlier generations in their reasons for moving abroad. In-depth interviews with young working Italians in Paris clearly show that finding a job is not their only concern. They talk about getting recognition and fleeing a country they unanimously describe as being unable to make room for young workers. Their narratives of injustice depict Italy as a country sunk deeply into both economic and moral crises, where a whole generation has no hope of social advancement.

This chapter contributes to the discussion on the relation between justice and mobility in showing how spatial mobility can be used as a response to a lived sense of generational injustice. It also shows that the experience of mobility contributes to shaping and strengthening such a 'sense of generation'. More broadly, the case of Italian migrants in Paris shows how a 'generational approach' can make a considerable contribution to research on international migration and mobility. It appears to be clear that in addition to work, the quest for social recognition plays a very important role in the development of migration projects. Italian migrants' sense of belonging is also connected to the use of institutional frameworks such as European mobility programmes as well as mobile media. Moreover, youth migration can be seen as the result of 'blocked' societies where transition to adulthood is experienced as too long by young people who cannot find their own rightful place. Spatial mobility is a response to weak or failed intergenerational social mobility, both in terms of economic position (access to better jobs and incomes) and social status or recognition.

Finally, a generational approach illuminates several debates in contemporary migration and mobility studies. By introducing 'generation' as a key variable, it may, for instance, help us to understand migration from an 'intersectional' perspective (Crenshaw, 1989; Kofman et al., 2011). The generational approach also highlights the internal diversity of the concept of 'highly skilled migration'. As higher educational attainment becomes commonplace, young Italian migrants do not describe themselves as a 'global élite', but rather as a new educated middle class spurred to emigrate from a country in crisis. A generational approach can thus also move forward the discussion on the 'middling' of skilled migration (Conradson and Latham, 2005; Mueller, 2013).

Notes

1 For a literature review on the issue of generation in migration, see Kofman et al., 2011.
2 Silvio Berlusconi is a businessman and a former Italian prime minister. He has been a key and controversial figure of Italian political life during the 1990s and 2000s. Most respondents refer to him as a symbol of Italy's moral crisis. Beppe Grillo is the leader of the Movimento 5 Stelle, a populist political party which achieved unexpected success in the 2013 parliamentary elections. This success was partly based on an extensive use of participative social media. Interestingly, Grillo's website is one of the most read blogs in the world.
3 'Economic miracle' is the commonly used term for the decades of great economic growth in Italy following the Second World War.

8   Sharing mobile space across generations

Lesley Murray and Susan Robertson

Introduction

With the proportion of people living in the world's cities set to rise to 70 per cent by 2050 it is becoming increasingly important to understand the ways in which urban areas accommodate people of different ages. Urban planners and designers have been grappling with ways to meet the challenges of accommodating diversity in today's cities, and 'shared space' (Engwitch 2005; Department for Transport 2007; Hamilton-Baillie 2008) has been advocated as one way of doing so. This planning and design concept is used to frame a new philosophy in street planning, which is based on all users taking responsibility for their impacts on other users. Such 'users' are mostly differentiated according to travel mode, but here we focus on the intergenerationality of shared space; with a critical approach to the intergenerational sharing of certain spaces and the co-construction of generational differences and space. In particular, generation, in close affiliation with age, becomes manifest through urban rhythms, and in particular rhythms of speed. Hence in this chapter we became rhythm analysts calling 'upon all [our] senses' (Lefebvre 2004, 31) and pursuing a necessarily interdisciplinary approach. The contention of this chapter is that particular spaces reveal intergenerational mobilities (Murray 2015) in ways that are often obscured, so that the mobility aspirations of particular groups of people become marginalised. One such space is 'shared space', an example of urban design that seeks to endure 'on people's terms' (Gehl 2010). 'Shared space' as a design tool created by Dutch transport engineer Hans Monderman, seeks to reconfigure street space, and in particular, street intersections, so that the semiotics of the space are disrupted. This in turn interrupts the usual practices of street space and the hierarchies within it. As the street is no longer marked out for cars to pass through in designated linear rhythms, and pedestrians to follow a code of compliance, the space becomes, to some extent, re-appropriated.

It is argued that at the heart of 'shared space' is a determination to slow down or freeze urban movement. In a 'hypermobile' society where people of particular ages are subjugated by the pursuit of speed, this unique space can reveal potential for improving the lives of people of all ages in cities. At the same time, there has been much criticism of this approach to street design, especially from cycling organisations, in relation to claims that 'shared space' represents a panacea for urban traffic problems. The DfT appraisal of the implementation of shared space schemes found that it impacted positively on traffic levels only if traffic flow and speed were below an optimum level. So on roads like Exhibition Road in London that have high volumes of traffic, the relatively high-cost improvements have been called into question. This is a valid critique in the context of a broader geopolitical context in which the notion of 'public space' is transforming. We are arguably amid a crisis in urban public space heralded by a retreat to private and semi-private spaces, which, Alves (2007) argues, is due to a weakening of the political dimension of the city and a redefinition of the notion of shared urban life. This, it is argued, is partly due to the prominence of the car, and spaces for the car in cities. But 'shared space' does take the car to task. It is the car, rather than the pedestrian that is potentially subjugated and there is hence potential to contribute to a critical investigation of the micro-politics of street spaces. Street spaces are usually set out in such a way as to privilege automobility and the embeddedness of a culture that gives priority to cars in urban space, where speed is privileged over slowness. This gives rise to the observation by John Urry at the beginning of his book that paved the way for a mobilisation of thinking in sociology that 'it seems as if all the world is on the move' (Urry 2007, 3). At the same time there is recognition that this increasing mobility is uneven, as full access is only possible by particular groups, including according to age. People are excluded from the available opportunities in street spaces according to age, as streets are designed to function as efficient thoroughfares rather than spaces of habitation.

In this chapter we engage with the ways in which these spaces offer opportunities for people of different ages to 'be' in urban street space. These are spaces that are, by definition, differentiated and in constant negotiation. However, this negotiation has become obscured as automobility creates quasi-private spaces of exclusion. Age here is considered from an intergenerational perspective, where we are all 'interdependent beings' who are always in a process of 'being' and 'becoming' (Uprichard 2008, 307). This allows a less linear focus on age, which is considered relational to other ages. In turn the relationality of age allows us to think about the ways in which rhythms of age overlap as rhythms of slowness and speed overlap and intersect in public space (Lefebvre 2004). It follows therefore, that a focus on speed can make embodied experiences of marginal groups more visible and this becomes possible in 'shared space' as opposed to spaces designed for speediness, where interactions become less distinct (Robertson 2007). 'Shared spaces' are spaces of slowness, or slowed-down-ness, and this can not only make visible different rhythms of movement but also lead to re-appropriations. As Jan Gehl, the designer of Brighton's 'shared space' attests: 'by creating a new type of street in the city and in the UK: Brighton now has England's first shared space street where cars are welcome – but on people's terms' (Gehl 2010). The concept of shared space becomes a means to re-appropriate space as well as to share it (Hamilton-Baillie 2008). By reconfiguring the materiality of these spaces, modal hierarchies, which discriminate according to age, are rewritten. Speed can make some embodied experience invisible. People who are marginalised from society, such as older and homeless people, are characterised by their lack of speed and so difference is understood as dependent on speed differential. Slowing down can make difference visible, as these differentials are open to scrutiny. Here, therefore we acknowledge the potential for an overdetermination (Sennett 2006) of urban space and focus on the social production of space (Lefebvre 1991) through the generative relations of the spatial and the social.

Differentiating space

The capacity to live with difference is, in my view, the coming question of the 21st century.

(Hall 1993, 361)

Before exploring these contentions with reference to a research study of 'shared space' we first consider why it is important to understand the ways in which different groups of people co-inhabit and negotiate shared space. Cities are characterised by difference (Young 1990; Massey 2005; Valentine 2008; Sennett 2006) and urban streets are the city stages in which this difference is played out. It is the way in which this difference is negotiated, produced and reproduced that in turn produces more or less uneven mobilities. Valentine (2008, 325) argues that previous attempts to consider new ways of being in cities, such as the 'cosmopolitan turn' (Amin 2006) have tended to romanticise urban encounters and that instead we should be working towards sites of 'meaningful contact' that last beyond the 'specifics of the moment'.

Space is differentiated according to age. This is because, as Vanderbeck (2007) argues, generations are situated in specific cultural contexts and will therefore experience spaces in different ways:

The emergence of contemporary forms of age segregation is linked historically to processes of economic and political change, including the growth of industrial capitalism that marked a shift away from home-based systems of production; challenges to the practice of child labour and the spread of compulsory schooling; and the creation of social security systems and welfare states.

(Vanderbeck 2007, 207)

Vanderbeck also suggests that intergenerational spaces, and streets in particular, can be positive for people of all ages, pointing to the negative impact of older people's withdrawal from streets on children's safety and security. He cites a number of projects that have sought to 're-engage' generations, including 'open community forums; participatory action research projects; oral history projects intended to foster empathy and understanding in younger people regarding the older members of their neighbourhoods; and programmes of mutual assistance and support' (211). However, there appear to be few programmes that attempt to address age segregation in public outdoor spaces.

Questions of age segregation cannot, however, be explored in isolation from other social characteristics and identities. In particular, it is important to note the ways in which identities are constructed and played out in public spaces through urban encounters. It is crucial also to consider the context of power dynamics in which street spaces are designed and in which urban confrontations are situated. As Valentine argues, we need to address issues of equality as well as diversity, and this can be achieved through mobility. The practice of moving through space is a political one. The 'spatial stories' of philosopher Michel de Certeau are told through mobile practices. While de Certeau's work looks at the proliferation of meanings, he is also looking for: 'myriads of almost invisible movements ... to become a "proper place for people"' (de Certeau 1984, 41) – both finding a place for individual practice but also within an overarching knowable system. In this, de Certeau may seem contradictory but he is drawing attention to the potential of 'thin' practical tactics within, or in spite of, 'fat' governing strategies. Mobile experiences are embodied and these are the experiences that produce mobile cultures and also contest and transform mobile cultures – being mobile is a political practice. In understanding the potential of the alternative space of 'shared space' we consider the ways in which shared spaces are indeed intergenerational and what this reveals about the potential for improving the lives of people of all ages in cities.

Slowing and revealing rhythms

Shared space is space in which movements are slowed, as automobile movements are slowed. We set out to capture the consequences of this slowing of movement using a methodological approach that encompassed the closeness of phenomenology (Merleau-Ponty 1962) and the distance of Lefebvre's 'rhythmanalysis' (2004). We asked what the slowing of movement in shared spaces reveals about the use of space by different generations. As well as acknowledging the slowing of the research setting, another key aspect of the research was the embodied experience of the space, which we explored through its visuality, with an 'understanding of images as meaningful objects central to symbolic and communicative activity that is core to many theorizations of contemporary visual culture' (Rose 2014, 9). However, although our epistemological focus is on the visual, following Pink (2009) this does not mean that we are privileging this sense over others but understanding that all the senses are interconnected. Instead we use this approach to reveal multisensory experience (Murray 2009) and to engage more specifically with the intersections between visual cultures and experiences and visual methods, where visual experience is embedded in social and cultural practice (Rose 2014).

We chose methods that spanned both the social sciences and arts, they were both sociological and architectural, and carried out an interpretative study, a 'bench study'[1] over a 24-hour period. This involved observing a particular bench in New Road and carrying out a small number of inter-views with the bench's settlers as well as observations of social interactions along the bench, and the documentation of these interactions using photo-graphs, video and drawings. It became clear from our initial analysis of this multi-media data that the street produced and was produced (Lefebvre 1991) through a number of intersecting and divergent rhythms: rhythms of order that occupy representations of space; the polyrhythms of spatial practice; and the subversive rhythms of representational space. As Lefebvre argues, everything – lived experience and representations – can be analysed as rhythms. 'The cyclical is social organisation manifesting itself. The linear is the daily grind, the routine, therefore the perpetual, made up of chance encounters' (ibid., 40).

Subversive rhythms are similarly present in Vergunst's study of rhythms of mobile embodied experiences in Union Street, Aberdeen, where mobile experiences did not 'correlate neatly to the structuring historical influences of its architecture and ultimately raise rather different questions of temporality' (Vergunst 2010, 380). Vergunst recognised the street level segregations:

> Those who feel in control now are the mobile, usually the young, the unburdened, and the skilful, who can cross the space in a minimum of time. For others, those laden with literal or metaphorical baggage, the street is a place of struggle, of time spent and lost.
>
> (Vergunst 2010, 381)

Lefebvre's rhythmanalysis offers a way to make sense of these street segregations and the dialectical intersections of governed space and experienced space. For Lefebvre (2004) rhythm originates in the body – in the rhythms of the body, the breath, the heartbeat. Cyclical rhythms, for example rhythms of the sea and temporal rhythms of monthly and daily cycles, intersect with linear rhythms. As rhythmanalysts we 'listen[ed] to the world, and above all to what are dis-dainfully called noises … and to *murmurs* … and finally [we] will listen to the silences' (ibid., 29). We observed the space from the equivalent of the window, taking up vantage points that surveilled the space from a distance, but not too far away, observing the fast and slow rhythms and the uses of time according to social categories and age.

We identified some of the rhythms that Lefebvre discusses: the 'measure' of the constituents of rhythms may be identified in drawings, photographs and video – for example through the digital stamping of time on photographs – but often we were searching for more nuanced indications of rhythm. We identified 'polyrhythmia' where the concurrent and simultaneous rhythms of the bench are linked to its materiality and meanings along with the 'natural' rhythms of the occupants whose bodies may be considered as terms of reference: the bench is something to sit or sleep on; it is also a refuge as well as a dirty old thing that attracts noise and trouble. Following Lefebvre's search for the 'hierarchy in this tangled mess, this scaffolding? A determining rhythm' (ibid., 43), the focus shifted to dominant rhythms, those that 'determine' the dynamics of the space at any one time. These dominant or 'staging' rhythms also change; for example, the police presence in the morning, to check on the homeless people, pervades the space. However, what becomes apparent, particularly in the evening, is that the usual dominant modal rhythms of the street are, in this space, irrelevant as the volume of people dampens the rhythms of auto-mobility during the day and is in direct conflict with pedestrians at night when young people appropriate the space. We also identified the 'arrhythmia' (abnormal rhythm) through a man making and selling origami and an older couple permeating the festivities at night. These practices stand out in contrast to the more pervasive rhythms. As Lefebvre observes:

> The relations of the cyclical and the linear – interactions, interferences, the domination of one over the other, or the rebellion of one against the other – are not simple: there is between them an antagonistic unity. They penetrate one another, but in an interminable struggle: sometimes compromise, sometimes disruption.
>
> (Ibid., 85)

In an analysis of the practice of skateboarding and its relationship to the rhythms of the city, Iain Borden (2001) stresses the connection between body, board and terrain and gives a detailed description of the 'micro-experience' of the sound and feel, the sensuous geography; the 'judder, hum and grinding' (ibid., 4), and the transmission of texture of the surface through the wheels and board to the body. The combination of clicks and roars and silence, in the air, make the soundtrack; the rhythm, added to the speed of walkers, acts as a counter rhythm. Borden vividly evokes the sensual experience of the instruments of rhythm. The practice of skateboarding, for example, can be seen as 'appropriated' activity that is 'in harmony with itself and with the world' (Lefebvre 2004, 85).

Cresswell points to the uneven analysis of mobilities across the quantifiable 'facts' of physical movement, the representations of movement that are often understood as metaphors, and last 'the experienced or embodied practice of movement' (Cresswell 2010, 19). These are notions that are very similar to Lefebvre's (1991) trialectics of space and Cresswell is calling here for a more holistic understanding of mobilities that requires 'paying [equal] attention to all three ... aspects' (Cresswell 2010, 19). The visibility and acknowledgement of various rhythms are differently understood: 'One person's speed is another person's slowness' (ibid., 21). In this study we are disentangling the inter-connections of individual movements that may slip in and out, cross through, disrupt or accentuate the multiple simultaneous rhythms of various scales. We seek to understand the impact of these negotiations in relation to age. One way to disentangle these is to consider the most readily documented rhythms through a range of recording techniques. The documentation gave us evidence of the dominant or 'determining' rhythms that we discuss later.

Intersecting rhythms: material, spatial and social

We set out to examine the relationship between generational experiences of shared space and the spatial and temporal rhythms in place. We looked at the repetitions of mobile (and settled) practices that rhythm necessarily involves, capturing Lefebvre's 'movements, gestures, action, situations, differences' (Lefebvre 2004, 25) over the 24-hour period, from seven o'clock in the morning until seven o'clock the next morning.

We observed that the material rhythms of the bench, of eating, entertainment, the weather and drinking, intersect with generational rhythms. The materiality of the bench produces particular generational practices, as one of the street users observed:

> Well because when you get to my age, you need something to prop you up otherwise you fall over, secondly, I'm not really ... I would sit out there, I mean usually these are full up so I just aim for a bench [...] But I would prefer a bench with a back to it because, you know, it's more comfortable to sit.
>
> (Sean)

The bench is sometimes used as a playground when parents bring their young children to climb over the structures, raising their eye level closer to that of the adults, so that there is a closing of the gap between the differences of bodily engagement across ages. The children are encouraged to clamber up and jump off, testing their physical prowess under the watchful eyes of parents:

> Well I have a – although I'm quite old – I have a small son and when he was two and three we used to come here to events like when the Queen came to Brighton we saw that, we saw that, and he would climb up on the top end of the bench when he was barely able to walk and wobble his way along and that, I remember that because I kept thinking he'd fall off.
>
> (Dave)

For some, the range of people using the bench is identified by their age and there is a positive reaction to the broadest range of ages:

> It's just quite a nice place, just lots of different types of ... there's like you see teenagers, elderly couples, families, so it's a real mix of people and it's

just a nice spot to kind of sit and watch the world go by and there's always something going on.

(Jan)

These material rhythms are productive of intergenerational space. Other rhythms were less episodic. Our observations showed that the space was very male over most of the 24-hour period. However, it was the cyclical rhythms, which are aligned to the rhythms of materiality: for example, to the rhythms of waste, which framed the intergenerationality of the space that we will now consider in more detail. We traced the generational rhythms through the day.

Morning

The first rhythms of the day were slow, remnants of the night before. As Lefebvre observes 'The night does not interrupt the diurnal rhythms but modifies them, and above all slows them down' (2004, 40). These early morning rhythms were slowed. They belonged to the homeless people as they gathered their belongings to temporarily leave the space before the arrival of the police, a daily event, and to the City Clean workers, clearing the bench of rubbish. By 8.00 a.m. the bench is completely clean and ready for another day. People move through the space, but intermittently: some service vehicles, a scooter, a skate-boarder, some cyclists and some pedestrians. At 8.30 it begins to rain and the bench is emptied. People move on either side of the street clutching umbrellas, keeping the middle line clear as if this is where the rain is heaviest. They gather underneath the balconies that line the west side of the street. The restaurants and cafés along the west side are setting up for the day and from 9.00 onwards there are more delivery vehicles. A family with two little girls walks along the bench, the girls climbing on top and holding their dad's hands. Perhaps the most poignant intergenerational moment is when a man in his twenties, merrily returning from his night out starts talking to an older woman, a stranger, who is coming back from an early appointment at the hospital. She is very pleased with the news she has just received from there and the man gives her a hug.

At this time of the day the homeless people are most visible, although conflict is hidden. The rhythms of homelessness – of mainly middle-aged and older people on the surface starkly contrasted with the rhythms of order through policing and control. Here we can identify at least two rhythms: the regularity of the policing and the uneven emotional responses of the occupants as their relationship to their home and their physical comfort is disrupted. However, they generated their own rhythms of order – clearing rubbish from the bench and making their home. The bench had its own rhythm of shifting materiality through the day and night: sometimes wet, damp or dry, intermittently laden, light, messy, tidy or dirty. The co-produced rhythms of the material bench and its occupants, whether humans, birds or dogs, could also be captured in the documentation of the rhythms of waste.

Figure 8.1 The bench in New Road, Brighton

The less visible rhythms of conflict were between businesses and those who are considered to upset the normative rhythms of the street.

> Well it can do, if we've got a very, at times I've had to get the community wardens to go and speak to a band that are playing here and all my customers are sitting here, 'I'm not listening to that racket', ... I think one long bench, I don't know, litter wise and people probably sleeping on there, I was under the illusion it would be designed in a way that would have not allowed the antisocial behaviour so much as happened.
>
> (Kate, business owner)

This may be the rhythms of ambient sound – the 'racket' of the street performers. Steve, one of the buskers, identifies this construction of 'us' and 'them', those with economic interests and others whose practices are considered a threat to these interests: the street performers, the street drinkers and the rough sleepers. The inequalities were revealed through comments from both perspectives:

> So that's one thing about this whole bench here, the sad thing is as soon as you sit here and talk to people that might look a little bit worse for wear, everybody [the people who own businesses along the street] judges you ... as soon as you sit with people who look like they maybe a little bit poor or homeless you get a reputation ...
>
> (Steve, busker)

This is played out in the movements of individuals from one place to another, or the micro scale mobilities along the bench may be measured as very similar. However, the significance of these movements is contingent on the relational engagement between occupants; that is to say that resting on the bench is considered very differently by other occupants depending on if the sleeper is identified as homeless or as a weary business person.

Afternoon

It is after midday when the street comes alive with people of all generations. There is a mixing, a porosity (Sennett 2006), a complex ballet of the street (Jacobs 1961). These aspects of the street space demonstrate the inseparability of time and space as integral markers of mobility experiences. The bench becomes the base for an upsurge in intergenerational interactions. When associated with 'family' this can sometimes be considered to represent the right to exclude others. For example, as one of the business owners remarked:

> I mean the sort of people everyone would like it on those benches are family people, normal people during the day having a lunch, throwing their rubbish away and then moving on, but that doesn't always happen, there's a lot of antisocial behaviour on the bench …
>
> (Kate, business owner)

There were calls for more interventions in the space – to control practices that were considered anti-social:

> You can't … that's the problem with uh public spaces, you can't make the benches too comfortable because people like to sleep on them so it's getting that balance right between being functional and not a bed for a tramp.
>
> (Shona)

In the middle of the day more people gathered in the space, people of different generations and multiple practices, as we have already heard from Jan: 'it's a real mix of people and it's just a nice spot to kind of sit and watch the world go by …' Differences were celebrated and considered to be a significant aspect of what makes the bench so appealing:

> I think it works, I think it all works very well, I like the fact that you have the restaurants there just outside, it creates kind of like a Parisian-type feel [...] A bit more exotic and it's quite exotic and these benches are very nice and I like sometimes there's buskers and things going on and the park is there, I think it's a very nice street.
>
> (Lisa)

There were high levels of intergenerational activity at particular times around particular activities: younger people at night, dancing, appropriating space, contesting their rights to occupy the space by stopping cars and climbing and dancing on them. The bench was at the centre of these activities, considered as a space of acceptance.

> Of course, you know, of course, it wouldn't matter about homeless people sitting on them, it wouldn't matter, it's part of the city anyway, but you know it would be good to have a lot more sitting spaces where you could view people, a few things going on, yeah that's what the city needs, more bench space [laughs].
>
> (Fred, sitting on the bench)

Some rhythms become established over a short time and attract others. This was seen, for example, with the performance of two young boys (aged 8– 10) dancing to music just in front of the bench in the early afternoon. Their performance attracted the attention in particular of others of a similar age, especially girls, who lingered on the benches to watch.

It is the interactions of people and space that create this 'porosity' (Sennett 2006) and the slowed-down-ness of the space seems to contribute to producing social interactions. There was also acknowledgment of the slowing down of cars in creating this relatively diverse space:

Figure 8.2 New Road in the afternoon

It's quiet today, so it's not bad, we've not had to dodge out of the way of any speeding people, so I think that whoever ... all the cars that are down here, they're um ... they understand that there's people as well. So if they're parked at the sides and um ... so there's plenty of room to still walk so they're considerate of each other, I guess ... We uh ... we live in Bedfordshire so we've come on the train but the person driving past now, slowly, knows that there's people around so I think if it works ... if it works well, it's a good idea if people respect that there's other people around, more transportation as well.

(Jim)

Night-time

At night the space becomes less diverse in terms of age. It becomes re-appropriated as a party space and this appears to deter the practices of lingering that predominate in the middle of the day: 'I just come out to play my drum at night so I'm just waiting night time to come and I start to do my drum, playing drums' (Bob). The rhythms that are co-produced between a less diverse generational occupation of the space characterise the mobilities more clearly; the dancing and sitting, drinking and interacting with other partygoers allows for a clearer set of differences to be seen in those who are either not involved or who are contesting the appropriation of the space in this way.

Figure 8.3 Dancing on New Road

I've been here only one and a half weeks, I come from and I spend loads of time here because I'm homeless for the moment ... Spend time here, the same people every day, you know [...] and especially nights and on the weekend when they play music, loads of shops like, I like those ...

(Maddy)

Yeah, so it is really good and it's really, it's really busy like the festival was really fun, there's always loads of stuff going on. Obviously we get the negative side because there's loads, there's like street drinkers and people hanging around like that so we've had a few like sick on our doorstep ...

(Jenny)

For some people, the space can become a place to avoid after a certain time, or there are micro spaces that become places to avoid, testing the tolerance of those who may share the street:

You know, yeah, you know when you are in life you mixing with a lot of people, some can be good, some can be bad but you know, it's hard to judge people [...] Yeah, I'm the type of person, I don't mind who you are or you alcoholic or whatever you are, I understand everyone have his own circumstances.

(Bob)

Levels of risk, tolerance and desire to be out late at night are factors of age and the behaviours that become more dominant, louder and more clearly visible at night appear to dissuade older people from spending time in New Road and children are less likely to be out in the street late at night; the diversity of generations at this time is considerably reduced and the space takes on an exclusionary atmosphere.

Conclusion

We have studied the ebbs and flows of the changing rhythms and interrelationships between generations, considering the material and the less visible rhythms and co-production of the space of New Road with particular focus on the occupation of the bench. We found that, at certain times, the dominance of one generation produces a diminishing of other generations and that the mobilities we observed and documented in the space are dependent on the spectrum of intergenerational occupation. Equally, we observed that the extent of diversity of generations occupying the space is dependent on the range of mobilities that are encouraged, allowed and tolerated in the street. Through the 24-hour period of our observations there were times when certain mobilities became re-enforced and, at these times, differences were more clearly visible and understood as, for instance, certain ages were effectively excluded at night time or certain mobilities were contested, such as when young adults playfully attempted to stop the taxis by jumping onto them. Through the slowing down of the 'shared space' we are able to identify difference more precisely and to consider how society and culture places people with differences such as homelessness, disability, gender and age.

Importantly, this study shows the ways in which, using Lefebvre's rhythmanalysis, the complex intersections of space and time construct spaces, social spaces. The design interventions in spaces like New Road are made meaningful though the social interactions that take place within them – the street is intergenerationally porous, but not at all times, and not in all aspects of the space. It is this that urban planners, policy-makers and designers need to take account of in intervening through schemes such as 'shared space'. At the same time, this intervention presents new opportunities to use mobile and multi-sensory methods that are not always useable in street spaces that are not slowed. When Jane Jacobs embarked on her street ethnography in the 1950s, she did so at a time when automobility was emerging as a dominant force in urban design. However, even Jacobs (1961) stopped at the 'sidewalk'. Streets in which the division of pavement and road are removed in an attempt to open up the space for mixing, also allow us to extend our space of research.

Note

1 Bench study was an idea that emerged from a conversation with Jim Mayor, the local authority project manager for the site.

9 Residential relocations between mobility cultures as key events during the lifecourse

Thomas Klinger

Introduction – a lifecourse perspective at the intersection of transport geography and mobilities research

There is an apparent trend throughout geography and the social sciences involving cross-sectional studies being increasingly complemented by longitudinal and lifecourse perspectives. This includes work undertaken within various subdisciplines such as migration studies, population geography, economic geography and travel behaviour research. This strand of research is motivated by the fact that recent social constellations and behaviour patterns predominantly originate from decisions and developments which have taken place in earlier periods of life (Beige and Axhausen 2012). This is especially true for highly habitualized activities such as everyday travel behaviour, as conceptionalized by the mobility biographies approach (Lanzendorf 2003; Scheiner 2007). In particular, decisions which attract considerable physical and/or cognitive resources are likely to have a long-lasting impact on day-to-day behaviour patterns. The material dimension of these long-term processes becomes apparent, for instance, in the purchase or disposal of a car or in changes in accessibilities after a residential relocation. However, less tangible aspects such as values and beliefs, formed by socialization and peer group influence, are also important for future behaviour. Thus, both objective and subjective conditions of everyday life are inscribed in a person's biography. Consequently, the lifecourse perspective aims to combine the personal characteristics of the individual with more structural and external conditions such as the household-related, socio-economic and cultural framework with which an individual is confronted throughout the lifecourse. This approach is in line with work calling for a more integrative analysis of individual mobility practices and their structural and socio-cultural foundation (Manderscheid 2014).

Therefore, the work presented here shows how the lifecourse intersects with external framings (Hopkins and Pain 2007) by focusing on the concepts of generation (Pain 2001), urban mobility cultures (Götz and Deffner 2009; Klinger et al. 2013) and, more specifically, on family and intergenerational aspects as well as on peer group and social norm influences. Empirically, this involves analysing the everyday mobility of people who recently relocated between cities representing different mobility cultures. These mobility cultures include both urban form and infrastructural conditions as well as the attitudes and mobility preferences of a city's population. Following this line of reasoning, residential relocations are understood as key biographical events (Müggenburg et al. 2015) which in reaction to the modified external circumstances increase the likelihood of a change of everyday behaviour, including travel patterns. Residential moves are often triggered by other life events such as the establishment of a partnership, the birth of a child or a job change.

Taking the biographical embeddedness of daily travel as a starting point, this chapter asks how and to what extent residential relocations influence the mode choice for daily trips. Furthermore, we want to consider how related life events, motives for moving and, furthermore, household attributes and inter-generational relations contribute to changes in daily travel behaviour after relocation. Last, we want to shed light on the question of how the experience of a new material and symbolic framework of everyday travel, conceptualized as urban mobility cultures, influences mode choice at the new place of residence. In this respect, we follow the suggestion of Doughty and Murray (2014: 17) that we need to understand better the 'ways in which local cultures of mobility intersect with lifecourse issues'. By combining concrete aspects of urban mobility such as mode choice with social norms influencing everyday travel, this work is located at the intersection between transport geography and mobilities research, an approach which representatives of both disciplines consider beneficial (Bissell et al. 2011; Shaw and Hesse 2010).

The remainder of the chapter is structured as follows. In the following section the mobility biographies approach is presented with a particular focus on indivi-dual and structural factors influencing travel behaviour throughout the lifecourse. In the third section we describe the sample we surveyed to answer the questions mentioned above. Consequently, we present selected results from the survey, focusing on the individual, intergenerational and cultural factors of mode choice change after moving house; these are then discussed in the subsequent section. The chapter closes with a conclusion and an outlook on future research.

Mobility biographies – travel behaviour research in a lifecourse perspective

In this section we aim to conceptualize in more detail the development of everyday travel during the lifecourse. We thus refer to the mobility biographies framework, followed by a discussion of external factors influencing travel behaviour throughout the lifecourse. In particular, we introduce generational aspects and the concept of urban mobility cultures into the debate.

Biographical continuity and discontinuity of individual travel behaviour

Lifecourse approaches have been introduced in numerous subdisciplines of human geography and social sciences, most prominently in migration studies

(Mulder and Wagner 1993), time geography (Frändberg 2008) and travel behaviour research (Clark et al. 2015; Oakil et al. 2014; van der Waerden et al. 2003). In the transport and mobilities literature, to which this chapter contributes, lifecourse-related approaches have been labelled as *mobility biographies* (Axhausen 2008; Lanzendorf 2003; Scheiner 2007). Most of these studies have in common that they capture individuals' lives by longitudinal concepts such as trajectories (Oakil 2013), careers (Lanzendorf 2003) or pathways (Frändberg 2008). Often these concepts are differentiated by different areas of everyday life such as work, housing, family life or travel behaviour (Lanzendorf 2003; Scheiner 2007). All of these spheres of day-to-day behaviour are usually characterized by a high level of habitualization. This means that many activities, for example, work- or family-related ones, are repeated on a daily or weekly basis and, to put it more precisely, that specific recurrent cues always trigger the same reaction (Matthies et al. 2006). An example might be, for instance, a commuter train approaching which causes a person to run to the platform, or the ringing of the school bell which prompts a student to unpack her/his sandwiches. Consequently, everyday behaviours implicate a high level of reliability and continuity, assuming that they take place within a stable context. This is not to say that under such relatively constant circumstances everyday behaviour either follows purely pragmatic rationalities or remains completely unchanged and 'lifeless'. Rather, it is even then pervaded by discursive influences and sensual and affective experiences (Doughty and Murray 2014: 4). However, behavioural adjustments are especially likely to occur when the contextual framework changes, both in terms of the material and spatial layout and in terms of the socio-cultural environment within which the activities take place. These contextual changes divide an individual's lifespan into sequences.

Consequently, these turning points within a biography attract particular attention in lifecourse-related research, since they define the framework for future behaviour alternatives. Depending on the characteristics of the new context, some behaviour options are easier to realize than others. Context changes and turning points that impact on an individual's behavioural orientations can be differentiated regarding their cause and their intentional character. The most common categories of such context changes are *critical incidents, external interventions* and *life events*.

- Critical incidents are particular good or bad experiences that are linked to a specific situation or activity. These associations influence the preference for specific behaviour options in the future, even if the objective contextual conditions remain the same (Lanzendorf 2010; Oakil et al. 2014).
- In contrast, interventions are planned and intentional, mostly initiated by political authorities (Cao et al. 2009: 361). They include regulatory measures like the introduction of a congestion charge as well as so-called soft instruments like awareness campaigns.
- Finally, life events are modifications of living arrangements, which are mostly a result of an intentional choice. They can be related to different

aspects of everyday life such as employment and career or the composition of the household a person lives in. Examples are graduating from
university, changing job, meeting a new partner or the birth of a child.
Furthermore, Müggenburg et al. (2015) distinguish between actual life
events and long-term mobility decisions such as the acquisition of a
driving licence or a residential relocation, although not without pointing
out that both categories are strongly interwoven.

In general, context changes are usually complex configurations consisting
of several incidents from the above categories. In particular, residential relocations are often the result of preceding life events. Most context changes are
not expressed by a single concrete incident, but are better described as a
period of transition, characterized by complex processes of anticipation, synchronization and adaptation. Examples are the anticipation of childbirth by
moving to more spacious accommodation beforehand (Schäfer et al. 2012: 81),
the synchronization of car use by different household members, for example,
by organizing work-related schedules after starting a new job so that driving
to work is combined with chauffeuring the children to school (Schwanen
2007). A typical adjustment process is a further relocation some time after
first arriving in a new city, based on more detailed knowledge of the new
surroundings (Axisa et al. 2012: 353). These examples illustrate that the
cause-effect-direction is not clearly defined. Life events might provoke behaviour
change, whereas established behavioural preferences and underlying beliefs
and attitudes can influence the timing and form of life events and related
decisions such as the destination choice after a residential relocation. In the
context of residential relocations this line of reasoning is known as residential
self-selection, which means that people choose their new place of residence
according to their lifestyle orientations and, besides others, their mobility
preferences such as good accessibility by car. The self-selection debate has
recently been linked to the lifecourse and mobility biographies literature
(Scheiner 2014; Zhang 2014).
   Some authors criticize the typology of events described above as too
restrictive and technical. They also suggest taking into account less obvious
occasions like, for instance, memorials, festivals or anniversaries, which for some
participants might trigger behavioural reactions. This is because these events and
their underlying symbolism can transmit social norms, which themselves influence behavioural dispositions (Bailey 2009: 3). This internalization of values
and meanings through a (life) event suggests taking a closer look at the context
in which an event takes place and the influences originating from it. Biographical life events are situations which disrupt previously established habits
and strategies of everyday life that no longer work because of changed contextual conditions such as new spatial or social arrangements, represented, for
example, by household configuration or a city community. This temporary
period of insecurity and disorientation leads to relative openness and the
deliberate perception of the alternative meanings and behavioural options

provided by the new context. In this view, perception and imitation categories, which are themselves socially and historically dependent, are the key processes linking context and individual behaviour throughout a biographical transition (Hareven 2000: 14–15).

External influences intersecting with travel behaviour throughout the lifecourse – the examples of intergenerational relations and urban mobility cultures

To understand the effect of contextual influences in more detail, it is useful to first differentiate contextual factors. In general, the range of social and political life is potentially relevant in a lifecourse perspective or, as Bailey (2009: 2) puts it, 'mobility biographies cross-cut spheres of production and social reproduction'. These intersections (Hopkins and Pain 2007) include encounters with all kinds of socio-demographic, socio-economic, family and peer group related, intra- and intergenerational as well as spatial, temporal and broader political and cultural differentiations. For the purpose of this chapter I will concentrate on family and intergenerational aspects as well as on peer group and socio-cultural influences.

Life events are often accompanied by a modification in household and family constellations. In most cases this involves a change in the number of persons living together, for example, if someone is born or dies, moves in or out. However, life events can also result in a rearrangement of existing configurations, for instance if a family member switches to a new job or school, starts a new hobby or gains a driving licence. All of these incidents might have rather practical effects, for example, for the harmonization and synchronization of time schedules, including time slots spent together and ones spent separately. This clearly involves behaviour adjustments, for example, regarding the usage of shared resources like a television, the bathroom or a car. Besides these rather pragmatic negotiations, changes in household constellations might also involve attitudinal and value-related discussions, which themselves include a behavioural component. This is the case if young parents judge public space and urban traffic in a different way than they did before the birth of their first child (Schwanen 2015: 7). This changed perspective potentially results in an adaptation of travel behaviour and mode choice.

These household-related (re)arrangements clearly involve an intergenerational component. Again, this entails very pragmatic strategies such as the efficient usage of common resources like the abovementioned integration of the children's trip to school into the parents' commuting trips (Schwanen 2007). However, it also includes comprehensive and robust socialization and imitation processes, for example, if one generation copies the attitudes and behavioural preferences of the preceding one (Döring et al. 2014: 175). Last, affective and symbolic aspects are relevant, for example, if the shared commute is perceived as a moment of intimacy between parents and children (Dowling 2000: 351).

These family- and household-related norms and behaviour settings are again embedded in broader cultural formations. Consequently, some authors point out that there are different parental cultures, for example where parents have particular concerns around road safety according to societal norms. Subsequently, I argue that the city level is not the only possible scale on which to conceptualize cultural difference in terms of mobility and transport-related characteristics, but it is a particularly adequate one. This is because cities and metropolitan regions tend to share common frameworks of everyday mobility arrangements. Therefore, people often judge mobility on a city level and identify certain profiles such as *transit metropolises* (Cervero 1998) or *bicycle capitals* (Vanoutrive 2015: 503).

I thus refer to the concept of urban mobility cultures (Deffner et al. 2006; Götz and Deffner 2009), which is based on the assumption that objective and subjective components of urban mobility are interconnected and interdependent. Hence, urban mobility cultures can be understood as an integrative framework incorporating the travel behaviour and underlying lifestyle orientations and mobility-related attitudes (Busch-Geertsema and Lanzendorf 2015; Lanzendorf 2002) of a city's inhabitants. Additionally, rather objective and structural components such as infrastructure (Haefeli 2005) and built environment (Cervero and Kockelman 1997) are included as the material extension of cultural priorities. Moreover, mobility- and city-related discourses (Cresswell 2010; Freudendal-Pedersen 2009) and urban transport policy (Bratzel 1999; Buehler 2011) are components of the concept of urban mobility cultures (Figure 9.1).

In order to understand the external factors, which intersect with personal mobility biographies in situations of contextual change like residential

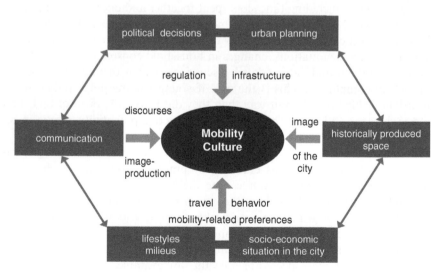

Figure 9.1 The concept of urban mobility cultures
Source: Deffner et al. 2006, own translation.

relocations, the concept of urban mobility cultures offers useful insights, since it serves as an integrative framework of individual travel behaviour. The potential transfer of these external factors into individual behaviour may work in many different ways. One is that the spatial layout of a city and the related transport infrastructure suggest the choice of a certain mode of transport. Moreover, the adaptation process might also be mediated by social norms and conventions, articulated by peers of the individual in question. This imitation of others' attitudes and behaviour has been labelled as *social spillover effects* (Goetzke 2008; Goetzke and Rave 2011).

In order to conceptualize these intersections between individual mobility biographies and contextual framings such as family and intergenerational constellations and also urban mobility cultures, we analysed the travel behaviour changes of people who moved between different mobility cultures. The empirical design will be presented in the next section.

Methodology – surveying long-distance movers in Germany

In an explorative study (Klinger et al. 2013) we categorized forty-four German cities by using a set of indicators representing the components of the 'urban mobility culture' concept (Deffner et al. 2006), namely objective and subjective aspects of urban mobility. By applying factor and cluster analysis we obtained six different groups of cities representing different mobility cultures: cycling cities, transit metropolises, auto-oriented cities, transit cities with multimodal potential, walking cities with multimodal potential and transit cities.

Starting from this pre-study, we intended to choose city-pairs representing different city clusters in order to survey people who had recently moved between contrasting urban mobility cultures. Besides this research-oriented consideration, we had to pay attention to pragmatic criteria such as availability of registration data and a sufficient number of people moving between the selected urban areas. We eventually collected data from people who from 2006 to April 2011 moved between the cities of Bremen (representing a 'cycling city'), Hamburg (a 'transit metropolis'), and the Ruhr area represented by the cities of Bochum (a 'transit city with multimodal potential'), Dortmund (not included in Klinger et al. 2013) and Essen (an 'auto-oriented city'). Even if the Ruhr cities were included in different clusters, all three are characterized by a rather strong auto-orientation (Klinger and Lanzendorf 2015). The selected case-study cities and their mobility cultures differ in many ways, for example in urban form, economic strength and modal split (see Table 9.1; for a more detailed characterization of the mobility culture clusters see Klinger et al. 2013).

In May and June 2011 we mailed out an eight-page questionnaire to 5,185 people, whose addresses we received from the municipal registration offices that collect both the current and previous postal addresses. The participants were asked and reminded once to send back the completed questionnaire. The number of returned questionnaires totalled 1,450, giving an overall response of 28.0 per cent of which 27.4 per cent provided valid data for the analysis.

Table 9.1 Case-study cities: socio-economic and transportation key data

City	Population size	Population density (persons/km²)	Household income (avg., €)	Modal share (%)				Mobility culture cluster (Klinger et al. 2013)
				Car	Transit	Cycling	Walking	
Hamburg	1,772,100	3,982	1,987	41	19	9	31	Transit metropolises
Bremen	547,360	2,891	1,784	41	12	16	29	Cycling cities
Bochum	378,596	3,940	1,466	53	14	4	28	Transit cities with multimodal potential
Dortmund	584,412	3,626	n.a.	50	22	10	18	–
Essen	579,759	4,208	1,539	54	12	1	33	Auto-oriented cities

The sample can be divided into six subgroups ranging from 121 to 296 people and representing the different city-relations that the residential relocations are based on (see Figure 9.2).

The comparison of our sample with the overall German population reveals that the participants of our survey are on average younger, better educated and less affluent. Moreover, they are more likely to be female, unmarried, employed and live in a single-person household. Our sample and the proportion of the German population which moved in the last five years show many similarities with regard to age, professional status and household size. Altogether these socio-demographic characteristics presumably indicate that students and young professionals are over-represented in our sample. This is confirmed by the relatively high level of education in combination with rather low incomes (for details see Klinger and Lanzendorf 2015).

From the questionnaire we derived variables measuring travel behaviour, socio-demographics (sex, education, income, employment), household composition (number of adults/children in the household before/after moving),

Figure 9.2 Study area in Northern Germany
Source: Authors' concept, map created by Elke Alban, Department of Human Geography, Frankfurt am Main University.

motives for moving and the perception of urban mobility cultures in both cities. Travel behaviour is measured by the frequency of car travel, rail transit (tramway, subway or light rail) and bicycle use, since these modes are considered as the backbones of the transport systems (Deffner et al. 2006) in the Ruhr area, Hamburg and Bremen, respectively. Subsequently, we created a categorical change variable, indicating whether mode use frequency decreased, continued on the same level or increased after the move, by simply comparing the frequencies before and after relocation. The comparative perception of the mobility cultures at the cities of origin and destination was captured by thirty-seven items derived from the basic elements of the mobility culture concept (urban form and infrastructure, travel behaviour and lifestyle patterns, transport policy and discourses) The items are divided into seven sections representing the main modes of urban transport (public transport, car, cycling, walking) as well as referring to travel behaviour, transport policy and transport-related media coverage. The items ask if certain characteristics apply more to the situation in one or the other city, based on a five-point scale ranging from 'in the city of origin' to 'in the city of destination'. Examples for items are 'Where are cyclists more accepted by other road users?' and 'Where is transport policy more advanced and innovative?' For an overview of all items applied please see Klinger and Lanzendorf (2015). The empirical data have been analysed with IBM SPSS Statistics 22.

Residential relocations between biographical and external influences – empirical findings

In the following section I will present empirical findings from our sample of people who recently moved between different urban mobility cultures. I will focus on biographical and generational as well as household-related and socio-cultural influences on everyday travel.

Age and motivations for moving, indicating the biographical and generational dimensions of everyday travel

As a first approach to understanding the biographical dimension of residential relocation and its impact on mode use change we divided the sample into four age groups: 18–29, 30–44, 45–59 and 60–74 years. As known from other relocation studies (for example, Scheiner and Holz-Rau 2013: 438), young age groups are overrepresented since they move comparatively often. Even if age-based conceptualizations of generation are highly debated (Hockey 2009), these 'lived categories' (Murray 2017: 4) can serve as a first approximation towards an understanding of generational behaviour patterns.

Cross tabulation and χ^2-tests for each of the four age groups and changes in car, rail-based public transport and bicycle usage after moving (compared to the sum of all other age groups) reveal some significant associations (Table 9.2).

Table 9.2 Changes in mode use after relocation in relation to age group (percentages)

Mode	Change	Age group				
		18–29	*30–44*	*45–59*	*60–74*	*ø*
Car use n = 1,310	Decreased	36.9	26.2	28.0	36.9	30.7
	Unchanged	34.5	41.6	55.1	47.7	40.6
	Increased	28.5	32.2	16.9	15.4	28.7
	*P value**	*.000*	*.000*	*.000*	.051	
PT (rail) n = 1,317	Decreased	31.7	40.8	29.7	31.1	36.1
	Unchanged	35.7	35.9	40.7	41.0	36.5
	Increased	32.6	23.3	29.7	27.9	27.4
	*P value**	*.004*	*.000*	.312	.676	
Bike use n = 1,331	Decreased	32.6	29.5	31.1	23.1	30.4
	Unchanged	39.6	39.9	47.9	47.7	40.9
	Increased	27.8	30.7	21.0	29.2	28.7
	*P value**	.434	.273	.117	.368	

Source: Own data and calculation.
Note: * According to the Bonferroni-correction the significance level is adapted to .0125.

For the 18- to 29-year-old movers driving decreases more than average, whereas the use of rail-based public transport increases more than average. Presumably this is due to a high proportion of students with relatively low income and car ownership levels. Furthermore, students in all five case-study cities benefit from so-called *Semestertickets*, relatively cheap season tickets for public transport, which they automatically receive with enrolment. 30- to 44-year-old survey participants show the contrary pattern, an above-average increase in car use and an above-average decrease in public transport use. This distribution points to the fact that for many university graduates the first job and a corresponding increase in income falls into this period of life, which makes a car more affordable. Furthermore, family formation might be a factor in using the car more often (Lanzendorf 2010). The mode use of the next age cohort, the 45- to 59-year-olds, is characterized by relatively high continuity, even after relocating and corresponding context changes. The proportion of unchanged mode use frequency is above average for all three analysed modes, even if this deviation becomes significant only for car use. This could be taken as a first indication of rather stable family and work arrangements, so that adaptations to everyday behaviour are not always perceived as necessary. Older people (60–74 years) included in the sample drive less – even if just not significantly – whereas rail transit and bike use are characterized by relative stability. This could be cautiously interpreted as a consequence of life events such as retirement and the related omission of work-related trips, which might result in a decrease in trip frequency and trip length and, subsequently, less car use (Hjorthol et al. 2010). Another factor may be, in some cases, diminishing physical fitness, although this relation is controversial, since car use might also have a compensatory function when physical limitations make walking and cycling more difficult (Haustein and Siren 2014). Taken as a proxy for generation-specific mode choice, these explorative findings indicate that mode choice after a residential move is more unstable and disrupted for the youthful and older people than for the middle-aged generations. In summary, it is striking that driving and public transport use are more affected by age than cycling is. This finding is in line with a more detailed analysis of the same data, which has shown that, in general, cycling shows relatively low reactions to socio-demographic variables (Klinger and Lanzendorf 2015).

Despite these insights regarding the link between age and mode use change after a residential relocation, an analysis restricted to age group effects remains simplistic and speculative regarding the biographical embeddedness of relocations, because of 'the situated, fluid and contested nature of age' (Hopkins and Pain 2007: 287). Therefore, to understand the causes of mode use change after relocation in more detail, a closer look at the motivations for moving house seems to be worthwhile (Table 9.3). Participants were asked to indicate if the relocation was motivated personally, by a job change, starting a course of study, or work for the first time. Further response categories – moving to a retirement home and purchasing a property to live in – were excluded from the analysis due to an insufficient number of cases. It was possible to indicate more than one motivation for relocating.

For all three modes only car use change shows a characteristic deviation from the average. This might be due to the heterogeneous character of this variable, presumably including situations as different as childbirth, moving together with or divorcing from a partner. Moving due to a change of job leads to an above-average persistence in car and bike use – even if not significantly for the latter – whereas changes in rail transit use do not show a specific pattern. This finding is, at least partly, in line with the results for the 45- to 59-year-old age group and can again be interpreted as an expression of relatively stable family and work arrangements. Movers who relocate to start a course of study show an above-average increase in public transport use and an above-average decrease in both car and bike use. This result corresponds to the behaviour of the youngest age group (18- to 29-year-olds) and therefore can be understood as a further indication of low income and a subsidized public transport season ticket for students influencing mode use change. Last, people who moved to start their first job drove significantly more but cycled and used public transport less than the average, a behaviour

Table 9.3 Changes in mode use after relocation in relation to motivations for moving (percentages)

Mode	Change	Motivation for moving				
		Personal reasons	Job change	Starting a course of study	First job	ø
Car use n = 1,337	Decreased	25.9	22.7	56.4	26.1	29.2
	Unchanged	41.0	48.7	25.2	35.6	40.4
	Increased	33.1	28.7	18.4	38.4	30.4
	P value	*.000*	*.000*	*.000*	*.004*	
PT (rail) n = 1,346	Decreased	37.9	39.1	22.9	40.9	36.5
	Unchanged	36.1	34.9	37.0	38.5	36.2
	Increased	26.0	26.1	40.1	20.7	27.3
	P value	.342	.225	*.000*	.058	
Bike use n = 1,360	Decreased	27.8	28.2	36.3	36.9	30.3
	Unchanged	43.2	44.1	35.8	38.3	41.8
	Increased	29.0	27.7	27.9	24.9	27.9
	P value	.068	.178	.103	.090	

Source: Own data and calculation.

pattern similar to that of people aged between 30 and 44 years. Again, it becomes apparent that use of the car and public transport is influenced by motives for moving and their socio-economic implications more clearly than bike use.

Changes in household composition and their impact on mode choice after relocation

The preceding analyses referred mainly to the individual level, even if age groups and motives for moving are strongly interwoven with contextual influences, as discussed above. Subsequently, external factors will be explicitly examined to explore their influence on mode use change on regular trips after residential relocation. Regarding modifications in household composition (Table 9.4), increases in the number of children younger than 18 years living in the household are of particular interest. Having more children in the household results in above-average increases of car use and above-average decreases in public transport use and cycling. Even if the changes in bike and public transport use are not significant, this finding, in

Table 9.4 Changes in mode use after relocation in relation to changes in the house-hold composition (percentages)

Mode	Change (n > 1,319)	Increase (adults)	Increase (children)	ø
		Household composition (change)		
Car use	Decreased	27.5	23.5	30.5
	Unchanged	37.7	39.9	40.6
	Increased	34.7	36.6	28.9
	P value	*.016*	*.044*	
PT (rail)	Decreased	37.2	40.5	36.0
	Unchanged	37.0	34.8	36.9
	Increased	25.8	24.7	27.2
	P value	.747	.454	
Bike use	Decreased	29.1	34.4	30.8
	Unchanged	40.8	38.0	40.9
	Increased	30.2	27.6	28.3
	P value	.567	.569	

Source: Own data and calculation.

combination with the results of the age-related analyses (Table 9.2), rein-forces the assumption that family formation and the related intergenera-tional commitment initiates a trade-off in favour of car use and to the disadvantage of more environmentally friendly modes of travel such as cycling and public transport. This confirms findings of multivariate analyses based on the same data, controlling for several variables such as spatial attributes, resi-dential preferences and cultural influences (Klinger and Lanzendorf 2015), as well as the work of others (Heine et al. 2001: 39–72; Johansson 2005). Further-more, an increase in adults per household leads to an increase in car use. This effect may be mediated by car ownership, given that the number of cars also rises in accordance with the number of people holding a driving licence within a household.

Residential locations as biographical transitions towards a new mobility culture

Perception has been identified as a key variable to capture external factors and their effect on behavioural change throughout a life event (Hareven 2000: 14–15). Consequently, a series of relocation studies have included attitudinal variables in order to capture the perception of, for instance, neighbourhood characteristics, travel mode satisfaction or accessibility aspects (Aditjandra et al. 2012; Cao et al. 2007; Chatman 2009; Handy et al. 2005; Scheiner and Holz-Rau 2013). In this study the comparative perception and evaluation of the urban mobility cultures related to the cities of origin and destination are of interest. Therefore, the items, referring to the perceived mobility cultures in both cities (see above), have been reduced to seven factors by principal component analysis. For details of the extracted factors please see Klinger and Lanzendorf (2015).

Table 9.5 illustrates the correlation between the perception of the seven components of urban mobility culture (columns) and the changes in mode use (rows). The percentages shown indicate the proportions of decreased, unchanged and increased mode use, when the individual perceived that the specific factor in question is more established in the city of destination.

At first glance, it becomes apparent that there are quite a few significant links between the perception of both mobility cultures and mode use change for day-to-day trips. This is a clear indication that both travel behaviour research and the lifecourse approach can benefit from interaction with cultural concepts such as the one used here. The increasing emergence of these concepts in transport and mobilities studies (Aldred and Jungnickel 2014; Hopkins and Stephenson 2014) is thus very welcome, especially consideration of social trends like the cycling boom in many Western cities (Lanzendorf and Busch-Geertsema 2014) or the less car-dependent current generation of young adults in urban areas (Hopkins and Stephenson 2014; Kuhnimhof et al. 2012).

Of the numerous aspects revealed by combining a biographical with a cultural approach, I want to focus on just two observations: First, car use seems to be most dependent on the cultural dimension of urban mobility. Although this impact diminishes slightly when controlling for other effects (Klinger and Lanzendorf 2015), it has interesting implications. Besides rather obvious linkages, for example, car use increases if the new mobility culture is perceived as being more car-oriented, it is remarkable that, first, driving also increases with the perception of more intense agglomeration effects such as more aggressive and dangerous traffic. This might indicate that some people feel safer when using a car. Second, bike use benefits not only from a higher cycling orientation, but also from increased walkability and a public-transport-friendly environment in the city the person moves to. This indicates potential synergy effects of the so-called 'green modes'. If a city, for instance, extends its public transport service, it is likely to be appreciated by cyclists as well.

Conclusion and outlook: residential relocations as the culmination of complex individual and external influences

This chapter has shown that travel behaviour research and mobility studies in general can benefit from enriching the mobility biographies approach with a conceptualization of the contextual effects an individual faces throughout the lifecourse. In order to do so, this work focused on residential relocations,

Table 9.5 Changes in mode use after relocation in relation to mobility culture perception (percentages)

Mode	Change	Urban Mobility Perception (Increase after move)							
		F1 –cycling orientation, environmentally trans-friendly transport policy	F2 – transit orientation, street life	F3 – walking orientation	F4 – car orientation	F5 – agglomeration effects, lack of safety	F6 – media coverage of transport issues	F7 – advanced transport policy	σ
Car use	Decreased	39.9	48.1	27.9	21.1	25.0	22.2	41.5	≈30
	Unchanged	40.9	33.3	42.2	41.0	38.1	41.2	34.1	≈40
	Increased	19.2	18.6	29.9	37.9	36.9	36.6	24.4	≈30
	P value	.000	.000	.481	.004	.000	.017	.001	
PT (rail)	Decreased	43.1	22.2	33.8	41.3	43.8	36.6	26.1	≈37.5
	Unchanged	34.9	35.7	33.8	34.4	33.3	35.3	46.3	≈35
	Increased	22.1	42.2	32.4	24.4	22.8	28.1	27.6	≈27
	P value	.244	.000	.618	.810	.003	.985	.002	
Bike use	Decreased	10.4	26.7	21.3	28.4	32.1	37.7	32.3	≈32
	Unchanged	34.9	40.1	44.0	43.2	39.4	36.4	44.4	≈40
	Increased	54.7	33.2	34.7	28.4	28.5	26.0	23.3	≈27.5
	P value	.000	.005	.003	.479	.958	.428	.269	

Source: Own data and calculation.

given that individuals are more open to external influences during these periods of complex contextual change. Accordingly, two contextual aspects, household composition and urban mobility cultures, have been discussed regarding their impact on the changes in travel behaviour of new residents. Both approaches have provided interesting insights into the adaptation and persistence of existing behaviour patterns after a residential relocation.

In particular, the concept of urban mobility cultures illustrates that it is promising to understand individual biographies as embedded in a broader socio-technical context. Consequently, it becomes possible to perceive travel behaviour not only as a demand, derived from spatial configurations and infrastructure supply, but also as a result of complex adaptation and imitation processes initiated by the local culture of mobility, for example, transmitted by peers with whom the individual interacts throughout this biographical transition (Goetzke 2008; Goetzke and Rave 2011).

However, the approach presented here also has several limitations. First, the described bivariate analyses ignore possible mediating and interaction effects among the various individual and external influences that affect travel behaviour change related to a residential relocation. Therefore, multivariate methods could enrich the analysis of the link between residential and everyday mobility patterns. Although this has been done by applying OLS regression models to the same data (Klinger and Lanzendorf 2015), the implementation of more sophisticated methods such as discrete choice or structural equation modelling is clearly recommended. Second, the process of the adaptation and internalization of contextual factors is still poorly understood. In this regard, qualitative approaches such as narrative interviews or participatory observation are promising for revealing typical decision and negotiation processes related to biographical transitions such as residential relocations. This is especially true insofar as quantitative approaches cannot fully capture the complexity of personal narratives, since respondents can only answer in pre-existing categories.

Nonetheless, it has become clear that quantitative analyses are not necessarily linked to a positivist epistemology, but indeed can be combined with a social constructivist research approach (Manderscheid 2014: 213–214; Wyly 2011: 904–905). In this respect, they are particularly helpful in order to reveal structural influences between social norms, discourses and collective constructions (Manderscheid 2014: 211) such as the lifecourse, generations and mobility cultures. This perspective might also be able to prevent an overestimation of individual potential as sometimes suggested by behavioural economics and, more generally, by a neoliberal world view.

10 Young age, mobility and social inclusion in a disadvantaged urban periphery in England

Miriam Ricci

Introduction

Scholarly interest in researching the physical mobility of young people and young adults has been comparably less extensive with regard to other age groups (Evans 2008), for example older people (Shergold and Parkhurst 2012) or young children (O'Brien et al. 2000). Sparse research engagement with this particular age group also applies to the academic discourse around transport, as a particular facet of the broad conceptualisation of mobility (Sheller and Urry 2006), and its interrelationship with social inclusion, defined as the process by which people participate as active agents in society, for example in decision-making, access to health and social care, and through participation in education, work and other valuable cultural and social activities (Kenyon et al. 2002; Social Exclusion Unit 2003).

In the past five decades, the assumption of mass car ownership and use, coupled with changing work and lifestyle patterns, has exacerbated conditions of disadvantage for the so-called 'vulnerable transport users' – low income groups, older and younger people, and those with no car access (Titheridge et al. 2014; Lucas 2004; Social Exclusion Unit 2003). Processes of social exclusion through lack of access or mobility have a clear generational dimension and vary considerably through the lifecourse, with people at the margins of the age spectrum being particularly vulnerable.

Despite enjoying a peak of interest in the late 1990s and early 2000s especially in the context of developed economies, the connections between mobility, transport and social exclusion have attracted less political attention in more recent years. Meanwhile, local transport policy, at least in the UK, has been slow in taking concrete action to systematically address the issues identified by the available empirical evidence on the links between transport and social exclusion (Lucas 2012). In this context, it is perhaps unsurprising that there is a relative paucity of peer-reviewed academic studies focusing specifically on the experiences of young people as users of the transport system and which examine the mobility contexts in which decisions about further or higher education, training and employment are made, and how participation in these activities takes place (for a study of mobility barriers to accessing higher education see Kenyon 2011). With this chapter I seek to address this gap and enhance the academic literature on young people's and young adults' physical mobility, with particular reference to the transport and social inclusion debate and from an intergenerational and lifecourse perspective. I provide a critical examination of the generational nature of mobility cultures, practices, needs and constraints emerging from the situated experiences of young age, and show how these can affect young people's life opportunities, choices and parti-cipation in further learning and employment. To this end, I will use selected key findings from a qualitative research study I carried out in 2014 with a cross-section of young people in Bristol, funded by the University of the West of England through an early career research grant scheme.

The rest of the chapter is structured as follows. The next section situates my research project in the broader academic discourse around the generational and lifecourse perspectives on young age, identity, mobility and social inclusion. I then illustrate the methodology I adopted in generating and analysing the qualitative data. Selected key findings are thematically presented and discussed by examining how social identity was represented in the context of place and institutionalised, adultist notions of young age; how mobility was discussed in relation to processes of growing up and 'moving on' through the lifecourse; what cultures, practices, needs and aspirations were attached to different mobilities and how these shaped, and in turn were shaped by, generational identities and intergenerational relations; finally, what constraints to inde-pendent mobility were discussed in relation to access to life chances and their implications for well-being. The concluding section suggests potential lines of academic inquiry for future research.

Youth, identity, mobility and social inclusion: an overview

Age and mobility have a complex interdependent relationship, recursively producing and reproducing one another in the lifecourse (Barker et al. 2009). Socially constructed concepts of childhood and youth reflect power relations between adults and younger age groups, normative expectations of what a child or a young person should be and do, and how much mobility, physical or virtual, they should be afforded. These expectations in turn shape the mobility experiences of children and young people, for example in relation to the extent of independent, unescorted travel they enjoy, which has significantly decreased over the past decades (Hillman et al. 1990).

In turn, limited control and independence in exploring and investigating the outdoor environment can prevent children and young people from developing the social skills they need to play an active role in community life, and can limit their visions and aspirations to become active citizens. Moreover, children and young people may also have distinctive preferences and aspirations in relation to experiencing place, which may differ from, but at the same be interdependent with, those of other age groups (Adams and Ingham 1998).

Place is key in shaping children and young people's social identities (Matthews 1992). A qualitative study of disadvantaged young adults found that wider frames of reference, self-perceptions and experiences were significantly rooted in and influenced by the local area (Johnston et al. 2000). The physical proximity to the neighbourhood, furthermore, determined whether education and employment opportunities would be considered viable, and networks of friends and leisure activities were all highly localised.

Over the past fifty years, remarkable social, economic and technological changes have radically changed the context in which young people make decisions about their lives, leading to physical mobility and access to transport becoming crucial to young people's social inclusion (Wixey et al. 2005; Jones 2012). Key services have been progressively located away from local communities and in areas with often limited public transport provision. Housing policy has also helped to reinforce this pattern, with poorer communities increasingly concentrated in isolated peripheral estates, so that vulnerable groups more likely to need key services are also least likely to be able to access them without a car (Power 2012).

The travel patterns of young people in the UK reflect such dramatic changes. Comparing National Travel Survey data over fifteen years, a study found that young people aged 17–20 have to travel much further for their most frequent journeys, such as to access education, work, shopping and to visit friends (Bourn 2013). Compared to other age groups, young people make the most bus journeys.

The saliency of transport in the daily lives of young people in the UK is highlighted by the Youth Select Committee's first inquiry in 2011, which found that safe, accessible and affordable transport was paramount to help young people access education, training or employment, and participate in society (British Youth Council and Youth Select Committee 2012). Cost was one of the biggest barriers, especially in the context of significant changes in education policy, with the withdrawal of the Education Maintenance Allowance scheme, and a steady increase in bus fares. Average bus fares in England increased by 33 per cent between 2007 and 2012, against a retail price index increase of 18 per cent (Bourn 2013).

Methodology

My research study focused on Lawrence Weston, a peripheral urban estate on the northwest fringe of Bristol, isolated from neighbouring communities by open space and motorways, and characterised by significant pockets of deprivation, lack of local employment opportunities, loss of local amenities (such as social services to vulnerable people, and a college site) and limited public transport connections to the city centre and key education and employment sites.

As part of the project, I conducted three focus groups with a total of eighteen young local residents aged 16–22, with the assistance of a local youth project (the BREAD Youth Project) and a community association (Ambition

Lawrence Weston). Participants included young people in employment as well as full-time and part-time students, apprentices, jobseekers and NEETs (Not in Education, Employment or Training). Additionally, I undertook in-depth interviews with seven professionals, referred to as 'key local informants', working and with relevant expertise on the target population and the study area. They worked in a variety of policy domains such as employment, skills and education, community regeneration, local economic development, transport and planning, and social work. The fieldwork took place between February and June 2014.

Focus groups are a popular research method used to understand people's views about specific topics, through examining the way participants express beliefs, attach meanings and provide their understanding of the world in the course of a facilitated small-group conversation (Wilkinson 1998; Smithson 2000). When carefully planned and delivered, focus groups are a particularly suitable tool to engage young people in social research, thanks to their informal nature and settings which can make young people feel more at ease than other qualitative face-to-face methods (Heath et al. 2009). Other methodologies are appropriate for conducting research with different generations. In-depth, one-to-one, face-to-face interviews were used for studying older people's mobilities (Musselwhite and Shergold 2013), whilst mobile and visual methods are particularly useful to explore the social, emotional and sensory responses of children and young people to mobile spaces (Murray 2009).

Prior to the fieldwork, I piloted the focus group schedule on a group of five female A-level students aged 17–18, resident in different areas of Bristol and recruited through personal contacts. Although their socio-economic backgrounds, life opportunities and aspirations differed from those of the Lawrence Weston participants, these young women provided valuable insights towards understanding the importance of place and culture and in shaping young people's identities, mobilities and access to life chances. Therefore, selected findings from the pilot focus group are included here. The research study was approved by the Research Ethics Committee of the Faculty of Environment and Technology, UWE.

The focus group process involved the completion of a short paper questionnaire survey, which collected general socio-demographic and travel behaviour information, in particular on the frequency and purpose of use of different transport modes. All interviews and focus group discussions were audio-recorded and transcribed verbatim for analysis with NVivo 10, a qualitative data analysis software widely used in social research. My analytical approach combined qualitative thematic analysis with the insights of conversation and discourse analysis (Silverman 2011). More specifically, I sought to understand how participants, through their utterances in group talk, framed the issues under consideration, what concepts and beliefs they brought into the debate and, importantly, how the interaction among focus group participants and between them and myself – the moderator – produced the various themes in the discussion.

Key findings

Young age, place and social identity

A key theme emerging from all discussions with young people concerned the concept of youth, its multiple meanings and attached social expectations, which varied across generations, and its interconnection to place. Furthermore, the groups highlighted how intergenerational relations, mediated through kinship and institutions, can at the same time generate and overcome access and mobility inequalities for different age groups and through the lifecourse. The term 'young people' was felt to be too generic and unable to capture the distinctive needs, aspirations, rights and responsibilities of 'young adults', as a few preferred to be called. This resonates with current academic debates challenging notions of childhood and young age as inherently transitional and individuals seen as 'adults in becoming' (Murray and Mand 2013).

A sense of powerlessness, unfairness and confusion pervaded some of the discussions around age and identity. A few participants questioned the rationale behind the multiple age thresholds at which children were expected to become adults in different policy realms, asking for example why they were considered too young to work (some employers and temporary work agencies only consider applicants aged 18 and over) or vote, but old enough to pay an adult ticket on the bus.

The inconsistency of different public policies in the UK, especially for young people over the age of 16, has been identified as a further problem in conceptualisations of youth, which leads to multiple and contested rationales for attributing responsibilities and rights to an age cohort that appears to be caught in a limbo between childhood and adulthood (Weller 2006).

In addition to questions around the meaning of young age, research participants raised issues concerning intergenerational justice by questioning why older people could get a free bus pass when they have a pension, while younger generations on no or low income have to pay a fare.

> I know this is really random, but why do old people get a bus pass? [...] Like, for [his] sister's situation, if she's not old enough to work, then why can't people like that have, I don't know, [...] maybe you can get on [the bus] for free if there's a good cause, if you know what I mean.
>
> (Focus group A, male, 17–19)

Furthermore, they challenged the prevailing social expectations attached to their age group and embedded in public transport policies, for example in relation to existing provisions for concessionary fares, timetables and routes. These, in their view, did not adequately acknowledge and provide for the wide range of education and work-related opportunities young adults may engage with (e.g. informal training, volunteering, part-time or occasional employment), but mainly cater for those engaged in traditional full time education and nine-to-five type of jobs, and along so-called 'key commuter routes' better served by public transport. Participants in Lawrence Weston spoke about the lack of public spaces available specifically to them in their neighbourhood, in particular indoor places 'open seven days a week that don't involve spending money', as one of the young women put it (Focus group B, 17–22). The lack of consideration for young people in the design and planning of public space has long been recognised (Frank 2006), so that young people often complain about having 'nowhere to go' and 'nothing to do' outside the home and school (Evans 2008). The Lawrence Weston groups for example noted that, while a much needed new playground for young children was being constructed as part of a neighbourhood regeneration programme, their local youth service had been reduced, and the groups expressed concerns about its future availability and quality. Evidence suggests that broader social and economic changes have progressively eroded the system of institutions with a specific remit to support young people, in particular those from disadvantaged backgrounds (Lawton et al. 2014). Finally, the estate as a place was significant in influencing some of the young residents' opportunities in life. These issues are examined in detail and in relation to mobility in the next section.

Mobility, growing up and 'moving on'

Both in the focus groups with young people and interviews with key local informants, physical mobility was framed as a constitutive element of well-being and processes of growing up, for example through the acquisition of the necessary skills and confidence to explore unfamiliar urban spaces. It is interesting to note that both older and younger research participants tended to reproduce adultist notions of childhood and youth, whereby the acquisition of certain sets of skills was seen as a key step in marking the transition to the adult stage of the lifecourse.

> It's really good for young people to almost, like, have to get the bus because it does give you some kind of, like, knowledge because when I asked my friends from my old school I was like, my house is literally on a bus route, you can't go wrong, and yet they were literally like, you need to give us directions to the bus, you need to tell us what to ask the bus driver [...] it kind of surprised me that they had no idea how to use public transport.
>
> (Pilot focus group, female, 17–19)

This connects with other research carried out with young bus passengers in London (Goodman et al. 2012) which highlighted the role of buses as places in their own right, sites for social encounters with people from all walks of life and different ages, and important tools enabling young people to explore and experience the city. Limited physical mobility was identified as a cause for concern and having negative effects on the local community's general well-being, and for young people in particular, in two different ways. On one hand, the progressive loss of local amenities and high concentration of disadvantaged families meant that Lawrence Weston was not attracting visitors from other areas or residents from a variety of socio-economic backgrounds, limiting the learning opportunities engendered by social diversity. On the other hand, the physical immobility of the young residents could in turn reinforce entrenched patterns of deprivation, and thus hinder processes of 'moving on' in life.

> Sometimes it's not an application form to fill in, sometimes it's looking in shop windows that they've got a vacancy here or a vacancy there, and you're not going to do that if you're not out and about.
>
> (Key local informant)

The extent to which the young research participants were physically mobile varied across the sample. By drawing upon the questionnaire responses and group discussions, the young female participants in the pilot focus group can be seen as multi-modal transport users, who were exposed to different modes in their daily lives and also confident and able to use them on their own. Their life opportunities and ambitions – they were all A-level students aspiring to go to university – seemed to be reflected in the way these young women framed their mobility and used transport. In contrast, the Lawrence Weston participants had been exposed to and were dependent on fewer transport modes than the pilot focus group women. Some of them admitted to rarely travelling out of the neighbourhood; a few had never taken the train in their lives, and most did not have a driving licence or access to a car.

Changing aspirations for automobility?

The stark difference in broader attitudes to transport modes between the pilot focus group and the Lawrence Weston participants is particularly evident when comparing and contrasting the opinions and travel behaviours of the two young female drivers in the pilot group and all the other young drivers in the Lawrence Weston groups. The female drivers in the pilot group were aware of and open-minded about other transport modes, especially public transport, which they used regularly to go to college. They were able to use the transport modes that most fitted their needs. They spoke of driving as an additional skill and tool, which would enhance their mobility opportunities in particular circumstances. In their talk, the car was represented as complementary to other modes, rather than as a substitute.

In contrast, all the young drivers in the Lawrence Weston groups were openly critical of other transport modes and considered driving the only realistic option that met their needs and allowed them to be in control of their time and, ultimately, 'free from the bus'.

> Rosie: 'As soon as I turned 17 I got my driving licence because I didn't want to rely on buses.'

Paul: 'Yes. Same. I couldn't rely on buses so it is sort of you've got to go through the pain of doing your driving licence and passing your test to actually be reliable, and if you're late at the end of the day it's obviously down to yourself and not buses.'

(Focus group A, 17–19)

Except for a few women, most of the Lawrence Weston research participants aspired to drive. Although it would be incorrect to regard this shared aspiration as a direct product of negative experience with the local bus service, it is nevertheless legitimate to argue that negative past experience with public transport does little to challenge widespread unfavourable perceptions of buses. In other words, negative experience contributes to reproducing and reinforcing the negative image of bus travel for young people and makes it even more straightforward for them to dismiss using public transport altogether. Pursuing car ownership and use, therefore, remains for this group of participants a powerful and unchallenged social norm.

I think now, like, previously, from using buses now I've got a car, I probably wouldn't ever catch a bus again because, like you say, they're just unreliable and, yes, I personally wouldn't step foot on a bus again to travel anywhere.

(Focus group A, male, 17–19)

Despite attracting generally positive comments, the car was framed by all participants more as a tool for enhancing personal mobility, rather than as a symbol of socio-economic status. All the young drivers, including those in the pilot focus group, readily acknowledged the costs and inconveniences associated with motor cars, in terms of getting a driving licence and acquiring, running, maintaining and parking a vehicle. Instead, those who aspired to drive but did not, tended to talk about cars almost exclusively in terms of their perceived benefits, such as independence and freedom, and overlook any potential disadvantages. In many (but not all) developed nations, such disadvantages have contributed to a decline in driving licence acquisition and car purchase by young people, especially by men in their twenties (Metz 2012; Delbosc and Curric 2013). Similar to the findings of Line et al. (2012), my research found that environmental considerations did not play a determinant role in young people's mobility decisions.

Another theme discussed in relation to personal mobility and driving was road safety, which was brought up almost exclusively and spontaneously by female participants, who expressed concerns with and fear of accidents. A few provided examples where they had been either witnesses of car accidents or one of the involved parties. Because of these concerns, several women claimed they would never want to drive and indicated the bus as their preferred transport mode.

Time constraints precluded the possibility to probe this issue more in depth during the focus groups. Nevertheless, several observations can be made on the possible factors underpinning this negative attitude to driving. While the road safety concerns explicitly voiced by the young women may indeed play a role, these may as well reflect broader self-esteem issues – for example in relation to the young women's perceived abilities to become competent drivers. Furthermore, negative perceptions of driving may also reflect wider beliefs about traditional gender roles and stereotypes, which might lead some of the female participants to represent driving as a typically masculine practice (Granié and Papafava 2011).

Young people as public transport users

The experience of using local buses was for most Lawrence Weston participants synonymous with inconvenience, loss of one's autonomy, and unpleasant experiences, which resonates with other studies of public discourse around bus travel, articulated predominantly as 'worst-case scenarios' (Guiver 2007).

> Young male 1: 'They're unreliable, they're never on time, they smell.'
> Young male 2: 'They never got change, that really gets me every morning. And they give you these change tickets.'
> Young male 3: 'You get proper weirdos on them.'
>
> (Focus group C, 16–20)

In one of the groups, after a long exchange about mostly negative experiences with public transport, a young woman summed up the overall low expectations of the local bus service with her conclusive remark 'Buses are like that' (Focus group A, 17–19). The interaction with bus drivers was a key theme in all discussions. Despite the pervasiveness of digital media in the young participants' lives, their engagement with public transport seemed to rely mostly on traditional means, such as timetables at the bus stop and, crucially, bus drivers, who were considered in all discussions as the key point of contact to get information on fares, routes and destinations. However, the nature of this engagement was discussed in negative terms, with most participants reporting instances where bus drivers had displayed unsafe driving behaviours and poor customer skills towards them and other passengers. In addition, the following quote exemplifies generational prejudices made explicit in the encounter between a research participant and the bus driver.

> When I was younger I was tall, as tall as I am now, and I used to ask for child tickets and they used to be quite rude to me about it.
>
> (Pilot focus group, female, 17–19)

The frustration with the quality and affordability of bus services needs to be considered in the context of broader public discontent with public transport in Bristol and associated significant events at the time of the study. Earlier in 2013, a campaign to improve public transport connections to education and other key services was launched by a group of young people in Lawrence Weston. At about the same time a bigger Bristol-wide campaign was carried out to persuade First Group, the largest bus operator in the area, to revise its fare structure. Both campaigns were successful in achieving some fare reductions, although these did not equally benefit all types of bus users in all neighbourhoods (Emanuel 2013). Affordability of bus travel was therefore still an issue for most of the Lawrence Weston participants, some of whom found it difficult to understand whether they were entitled to young people's fares and how they could claim them. My research findings resonate with the results of a qualitative study of young bus users across England (Passenger Focus 2013) which found that young people felt unwelcomed and undervalued as public transport users. Affordability and availability of bus services, especially off-peak, have long been identified by young people as key problems in using the bus to access life chances (Taylor et al. 2007).

Young people's representations of active mobility

The bus appeared to be the default mode of transport for the young Lawrence Weston participants without a driving licence, especially for the young women, who sometimes used the bus for very short journeys. Limited levels of physical exercise, also in the form of active travel, were identified as a key concern by an interviewee, who discussed the difficulty of promoting active and healthy lifestyles to local young people in the context of cheap fast food and takeaway culture and lack of affordable fresh food in the neighbourhood, combined with entrenched material and cultural deprivation:

> They struggle to get off the sofa sometimes in terms of their motivation and self-esteem and aspirations. You know they are fighting everyday things and sometimes mental health issues or feeding their families or sadly you know there's other issues, more serious issues just about basic survival going on.
>
> (Key local informant)

During the focus group conversations, walking and cycling were almost unanimously framed as leisure and social activities rather than means of transport, which is resonant with the findings from other studies of young people's mobility (Jones et al. 2012). While the majority of Lawrence Weston participants claimed to be able to ride a bike, not all had access to a bicycle and only few cycled regularly. Concerning barriers to cycling, young people mentioned issues around the image of cycling and practical aspects associated with cycling for transport. These included fear of traffic, the perception that bicycles break down easily, lack of confidence in riding a bike, concerns with cycling in adverse weather, icy roads or in the dark, the cost associated with purchasing and maintaining a good-quality bike, lack of safe storage space at home and lack of cycle repair shops in the neighbourhood.

One of my interviewees – a regular cyclist – suggested poor health as an additional reason behind the low levels of cycling among the local youth.

> Why don't people cycle? One of the biggest issues is around numbers of people who are overweight and obese. And we are probably worse than I have seen it anywhere. We have had young women that we work with go on a girls' project with other local young women from like Southmead and Henbury, and [health] workers make comments on how big the young women are.
>
> (Key local informant)

This latter comment mirrors data on health indicators published in the most recent Neighbourhood Partnership Profile report (Bristol City Council 2013) suggesting that levels of obesity in the Kingweston ward, which includes Lawrence Weston, are higher than the city average (with 62 per cent of adults overweight and obese compared to 50 per cent in Bristol).

As in the case of cycling, walking was rarely framed by the young participants as a means of transport and instead usually discussed initially as a leisure activity. As such, walking for leisure was associated with, and constitutive of, positive experiences and feelings of pleasantness, being close to nature, and spending quality time with family and friends.

Although a few Lawrence Weston participants claimed to rarely walk for transport, especially the young drivers or those regularly carried as passengers by others, overall walking was seen as a taken-for-granted activity that everyone undertook to some degree but to which most young people did not explicitly attach much importance. Although the health and financial benefits of active travel were acknowledged by most participants, walking and cycling for transport in the form of active commuting to and from work, were framed as markers of disadvantage – a manifestation, for example, of young people's inability to access or afford other transport modes and their exposure to safety risks from traffic, rather than a travel practice reflecting their personal preference.

> Where I work, I don't think there's any bus route. So, before I had my car I was getting a bike every day. I did that for about a year and that was horrible and no bus routes or anything. So, it's like you have to ride in all sorts of weather and stuff like that. It isn't very nice.
>
> (Focus group A, male, 17–19)

Constraints on young people's independent mobility and access to life chances

Although in principle all the young people engaged in this research had been able since their early teens to be physically mobile, independently of adults, through active or motorised transport, in practice they (had) faced a number of limitations to translating their freedom of movement into action. This had significant impacts on their overall freedom to access life chances. As one of the young men in a group indicated in relation to taking the bus to an unfamiliar area, travelling by bus may require knowledge and confidence that some young people do not feel they possess. The distinctive challenges faced by young people from disadvantaged backgrounds were further illustrated by two interviewees, who suggested that broader issues relating to learning difficulties, vulnerable family situations and low self-esteem negatively affected local young people's abilities to undertake bus journeys autonomously or to engage in other activities that most of us take for granted.

The affordability, reliability and availability of bus services to the desired destination and at the desired time of travel, for example to undertake work shifts outside of the core nine-to-five period, sometimes determined whether young people could actually travel by bus at all, and limited their access to life chances such as education, training and jobs.

> You know, obviously you're trying to being young, trying to get your first jobs and stuff to, like, move on and stuff, but you can be very limited on what jobs you go for. Say you wanted a job in town, sometimes the bus fare, by the time you get your bus fare there and back, say you're only working a four-hour shift, it's not worth it – you're going to make next to nothing on it.
>
> (Focus group A, male, 17–19)

This meant that many young people had to depend on lifts from friends, colleagues or family members, for essential and time-sensitive journeys. The following quote shows how intergenerational relations can be crucial to overcome mobility barriers for young people.

> The buses don't run down to where I work and I can't walk down there because it's dangerous. My work is literally around the corner. It takes four minutes to get there. So if it weren't for my dad giving me a lift, I wouldn't have that job.
>
> (Focus group A, female, 17–19)

The practice of sharing journeys had two competing representations. On one hand it was seen as a positive social practice, constitutive of relations of kinship/friendship and providing informal financial support, as getting a lift often involved making a contribution to fuel cost.

> We have, like, a big friends group and we don't all need to have a car; we all have a shared car. Like, Sarah, Sarah can, like, lift share and she doesn't really need a car.
>
> (Focus group A, female, 17–19)

On the other hand, the precarious nature of getting a lift for essential journeys like commuting meant that not being able to be autonomously mobile, thus having to rely on other people for transport, could also have negative implications, as the following two quotes illustrate.

> Sometimes there's not always [another transport option] so you're panicking and worrying and you put other people out of what they're doing, because they need to take you.
>
> (Focus group A, female, 17–19)

> My first job, I was working in Reading, and I was going, like, quite far away, but I couldn't drive or anything like that and I had to rely on a friend to take me who was working there as well. But then he got moved around just as much as I did so then I couldn't rely on him to take me so I couldn't go there anymore.
>
> (Focus group A, male, 17–19)

Conclusion and future research

By drawing upon data generated through a qualitative methodology and using an intergenerational and lifecourse perspective, this chapter has examined how competing notions of age and youth intersect with and co-produce the mobility cultures, practices, aspirations, needs and constraints of a cross-section of young adults in Bristol. Several key observations can be made to summarise the research findings. First, the experience and identity of young people as social agents is influenced by competing notions of 'young' age used in the public and policy discourses. These typically adultist conceptualisations of youth help to create a sense of powerlessness and unfairness among participants. This was further compounded by the perceived lack of agency to influence broader policy decisions affecting important aspects of young people's lives, for example in relation to delivering local youth services, planning public spaces and managing transport systems. Intergenerational relations were found to be crucial in creating and at the same time overcoming access and mobility inequalities for young people. Second, physical mobility was perceived as an essential factor underpinning the overall personal development and well-being of young people. By being mobile, alone or with friends and family, and through different means of transport, active and motorised, young people can develop the skills and confidence to explore the city beyond their comfort zone. In other words, physical mobility was seen as constitutive of processes of growing up as well as acquiring social agency.

Third, exposure to and positive experiences with a variety of means of transport other than the motor car from a young age may contribute to developing an open-minded outlook on all transport modes, which in turn may help prevent locking young people into unsustainable travel patterns, in particular solo driving. Growing evidence suggests that transport behaviours are the outcome of a complex interplay between person-specific and context-specific factors, and can change over the lifecourse as all these multiple factors change as well. Nevertheless, beliefs, attitudes and past experience can all play an important mediating role in the travel decision-making process, especially when transport behaviour is consciously and actively considered, rather than simply habitually performed, during a life transition such as changing school, getting a job and moving house (Chatterjee et al. 2013). These considerations are especially relevant for policies directed at young adults who are starting to make key decisions about their future, in particular concerning whether and where they will pursue further learning, or seek employment. At some level these decisions include, explicitly or implicitly, questions about transport.

Fourth, and in connection with the previous point, consistently negative experiences with specific transport modes do little to challenge cultural pre-conceptions and prevailing social norms. Public transport, and bus travel in particular, was overwhelmingly represented as unattractive, especially by those who did not, for a variety of reasons, have other options. Car driving, in contrast, was predominantly framed as the key to freedom and autonomy despite its perceived shortcomings, chiefly the associated costs and personal inconvenience caused by traffic congestion and parking. However, this was not the case for the young female drivers from more fortunate backgrounds, who were not (culturally or behaviourally) locked into any particular mode but willing and able to be mobile through a variety of means of transport.

This leads to the fifth point, which is about the importance of place, culture and socio-economic position in understanding mobility cultures and practices. My research found a stark contrast in the mobility experiences and representations expressed by research participants from different cultural, social and economic backgrounds. Very limited mobility, especially by active travel, can reinforce patterns of deprivation and poor health, especially for those young people who have few reasons, or opportunities, to leave the house. This is compounded by evidence suggesting that health inequalities are socially and economically patterned (Marmot et al. 2010). Finally, conditions of mobility disadvantage, along the dimensions of transport availability, accessibility, affordability and acceptability, coupled with situations of material and cultural deprivation, can produce two significant limitations. First, they constrain young people's freedom to choose and pursue their objectives in life, such as attending college courses and securing a job. Second, they influence the processes by which young people achieve their objectives.

Taking all these threads together, a few suggestions can be made in relation to possible future research directions. Concerning specific topics for further research, I can identify two possible avenues that could be important in advancing academic knowledge in relation to intergenerational mobilities and the lifecourse, and also to inform policy making. One possible research direction is to explore how the concepts of justice, fairness and entitlements intersect with intergenerational mobilities, transport systems and policy, which emerged as key (and largely unprompted) themes from all discussions with research participants. Further studies could explore how issues around intergenerational justice and competing notions of 'justice' are articulated by different age groups in discussions about mobility. Because of the importance of social inequalities in the construction of youth identities and mobilities, a further strand of research could address the ways in which social inequalities, including those reproduced through intergenerational cycles of deprivation, affect people in different phases of their childhood and youth, and how these can be mitigated.

Concerning research methodologies, I believe there is a need for more parti-cipatory types of research to open up debates around local policies, such as those affecting public and active transport, to local young people, especially those from more disadvantaged communities. If delivered through collaboration with the relevant local authorities, transport decision-making bodies and operators, participatory research could enhance existing public engagement processes, for example those undertaken within local 'accessibility planning' in the statutory transport plans developed by English local authorities. This process would help bring to the fore the (often unspoken) values and visions embedded in policy making, and place them in dialogue with research parti-cipants' own values and needs. Although the challenges associated with con-ducting meaningful, transparent and inclusive participatory types of research in transport and urban planning are well known (Hodgson and Turner 2003), the exercise itself could bring invaluable benefits in terms of increased mutual trust and understanding between all the involved parties.

11 Reliance mobilities

Christian E. Fisker

Introduction

The mobilities turn has opened up a view of the complex relationships between mobilities and immobilities. One can see the relational nature of mobility and immobility in the works of Adey (2006 and 2007), Classen (2006) and Goffman (1961), where some are relatively mobile while others are relatively immobile. One can also look to the relational nature of those attempting mobility, in their relationships with technologies and infrastructures, where some configurations are possible, while others are highly constrained or not possible, depending on the unique circumstances of an individual and those they may attempt to team up with in order to be on the move together. One can go still further and consider the relational nature of mobilities required to address human needs, in the relationships people attempt to create, configure and reconfigure; where they team up with others, such as family and friends, in various ways, where one person within the configuration is immobile.

This chapter is an investigation into the reliance mobilities of those who lived immobile lives within iron lungs, and those who interacted with them such as family, friends and health care workers, in order to address needs. Needs as defined in this chapter include physical and social needs; for more background on attempting to connect with needs please consider Fisker (2011a) and Chapin (1974). The primary sources for this investigation are four books written by and about people who lived parts of their lives in iron lungs, with particular emphasis on the accounts of Mimi Rudulph (1984). The chapter begins by setting out the relational nature of mobilities, followed by insights into iron lung technology. The chapter then examines accounts of life in and around those living in iron lungs and concludes with a discussion on how this inquiry may be of benefit to those researching and planning housing, health care, social services and transportation systems for housebound seniors. This chapter contributes to mobilities turn theory by proposing a means of thinking about mobilities that are created, often with an intergenerational nature, configured and reconfigured in order to address the needs of those who are immobile and by introducing 'reliance mobilities' to the mobilities turn vocabulary.

A relational thing

We can take our first steps into this investigation by considering the relational nature of mobilities. Long before the mobilities turn and considering mobilities within a particular setting, Goffman (1961) considered what he described as social establishments where there are 'fixed members' who provide a service and also a continuous flow of others who receive the service. Within the mobilities turn, various scholars, such as Adey (2006; 2007) and Classen (2006), have turned their attention to airports as settings that exhibit both mobilities and immobilities. Urry (2007, 54) argues that aeroplane technology, which is central to contemporary global experiences, 'requires the largest and most extensive immobility, the airport-city with tens of thousands of workers, helping to orchestrate millions of daily journeys by air'. Using Goffman, we can see airports in a sense being similar to a train station, where 'fixed members' are providing a service to a 'continuous flow' of others, where serving the needs of those who are mobile is given prominence. Expressing a relational nature of mobilities, yet giving prominence to mobility, as opposed to immobility, Beckmann (2005, 84) states that: 'mobility relies on immobility; it is precisely because certain subjects and objects are immobilized that others can travel'.

Kesselring (2008, 85) finds that '[t]he mobility of the one is the flexibility and the immobility of the others'. Cresswell (2006, 15) argues that tourists depend on the relative immobility of those who service the needs of travellers. Often the discussion of immobility seems to focus, or give prominence or foreground, the mobile in any relational discussions. Bissell and Fuller have recently noted that:

> a relational approach recognizes the unequal 'power geometries' that are such a significant part of mobility systems whereby, put very simply, the speed of some comes at the expense of the stillness of others.
>
> (Bissell and Fuller 2011, 4)

In the same Goffman (1961) work mentioned earlier, he describes the mental institution as a form of total institution, which has an encompassing nature symbolized by the barriers to social intercourse with the outside. Some residents are highly constrained within the setting; others are able to move around within the setting to some degree and still others are able to make their way outside the walls from time to time. At the same time there are employees who make their way to and from this setting in order to address the needs of the residents. In this case the relation between those who are relatively mobile and immobile is the mirror image of settings such as airports or train stations in the sense that those who are mobile are predominantly on the move in order to serve the needs of those who are relatively immobile.

Adey argues that mobility is a 'relational thing':

> mobility and immobility are profoundly relational and experiential. The point I am trying to make is that while everything might be mobile,

mobilities are very different, and they also relate and interact with one another in many different ways. This relatedness impacts upon what mobilities mean and how they work. It is also because of this difference and relatedness that illusions of mobility and immobility are created.

(Adey 2006, 83)

Now we come closer to a way of examining, and thinking about, the lives of those who lived in iron lungs who can be seen as immobile, or relatively immobile, and, as will be seen later, generating mobilities to support their immobile life. Beckmann describes immobility as 'the inability to overcome space – and mobility is the ability to do just that' (Beckmann 2005, 86). If seen from the vantage point of independent, unrelated mobilities, Beckmann may have a point. However, if one considers that someone can build up relationships with, and rely on mobile others (Fisker 2011a) to help overcome space, then indirectly, immobility is overcoming space. Already you've seen me use the term 'relative immobility' and also refer to Adey who uses such a term. As Adey states: 'there is never any absolute immobility, but only mobilities which we mistake for immobility, what could be called relative immobilities' (Adey 2006, 83).

Cresswell (2006, 166) strongly critiques Ivan Illich's suggestion that 'men [sic] are born almost equally mobile' as absurd, by noting the immobility of babies. I have argued that mobility and immobility is ever changing across the life course, with a higher reliance on others for mobility in our youth and later years of life (Fisker 2011b).

Adey notes:

> Some people are dependent on other people in order to move; children might have to travel with their parents or a responsible adult; a mobility-impaired person may be dependent upon a friend or a relative to help him or her get around.
>
> (Adey 2009, 23)

Here we see examples of what Jensen (2010; 2013) has described as 'mobile with' arrangements where two or more people team up and are mobile together. Another example is Nijs and Daems (2012) describing how an elderly person in their study is always accompanied by someone in what they describe as a 'collective body' where there is an escort for outdoor mobility; there is some form of teaming up in order to cross space together. In both of these examples the dependency appears to run solely in one direction, where the required skills to bring about motility (Kaufmann 2002) may rest with one of the two or more people in a team. Anecdotally, I am aware of situations where it is the combined skills of two elderly people that enable motility and actualized mobility. What is not described in these particular examples is where a relatively immobile person is dependent on the mobility of someone else, who delivers needs to them. Such a scenario is what I have described elsewhere as the 'mobile other' (Fisker 2011a; 2011b), which can include, for example, a family member or friend picking up groceries and bringing them to a housebound senior. Referring again to Goffman (1961), here we see a relatively immobile person orchestrating a mobile other arrangement at a mental health facility:

> I once saw a locked-in patient employ the standard device of dropping some money in a paper bag out the window to a paroled friend below. On instruction, the friend took the money to the canteen, bought some potato chips and coffee, and took these in a bag to a ground-level screened window through which the originator's girl friend was able to reach them.
>
> (Goffman 1961, 285)

Here the relatively immobile person is *orchestrating* the mobilities of a friend and girlfriend, acting as mobile others, in order to address a need for potato chips and a coffee. Another example of a mobile other arrangement is Okie's (2008) description of a nurse practitioner visiting a woman in her eighties, as part of a programme that addresses the needs of 500 elderly or disabled people in Boston who cannot get to a doctor's office, thereby reducing hospitalization and emergency room visits. Joseph and Hallman (1998) find that distances, measured as time travel, have a discernible effect on the amount and frequency of assistance employed by paid family caregivers to their elderly relatives, where as distances increase, care levels decrease. In her research on home care, with accounts from home support workers, elderly clients and family members of clients, Martin-Matthews (2007) notes the importance of considering the 'other players' in the home care dynamic. Describing his own late-life immobility, being housebound and the world now coming to him, Schwartz (1996) states:

> Support systems are essential when you're in a state of disrepair. I am lucky to have a whole stream of friends coming through my house. I call them my support community, my angels, my dear friends. They come quite regularly to find out how I am, to exchange thoughts about spiritual issues, to let me know how much they care. Sometimes they bring dinner. They come to have dinner with me, to communicate about the news of the day or what's happening in their lives, to tell me about issues they're struggling with, things about which I might be able to offer help or advice.
>
> As a matter of fact, there is a great deal of interchange, my giving to them and their giving to me. They tell me they are learning from me, that watching me is an inspiration to them. And in return I feel that they're continuing to keep me alive because there's so much energy and good feeling, love, concern, and care that comes from these friends, as well as from my family. Since I'm so restricted in my movements, they bring the

world in. They bring themselves in. By their bringing the world in, I can get outside to some degree.

(Schwartz 1996, 82–83)

While Schwartz, a sociologist, is considerably immobile in his home when he writes this, friends are altering their mobility in order to come to him, as mobile others. At the same time, Schwartz's mobile visitors are 'bringing the world in'. Mitch Albom, as a young man, wrote a book about his regular long-distance visits, travelling from Detroit to Boston to be with Schwartz while he was housebound. Here is a passage from Albom's (1997) vantage point as it relates to these encounters:

The last class of my old professor's life took place once a week in his house, by a window in the study where he could watch a small hibiscus plant shed its pink leaves. The class met on Tuesdays. It began after breakfast. The subject was The Meaning of Life. It was taught with experience ... You were also required to perform physical tasks now and then, such as lifting the professor's head to a comfortable spot on the pillow or placing his glasses on the bridge of his nose. Kissing him good-bye earned you extra credit.

(Albom 1997, 1)

Here we can see two distinct vantage points of these encounters: one from the elderly person's perspective and the other from a young adult's perspective.

Iron lung technology

Maxwell (1986, 3) argues that the iron lung represents medical technology in its most palliative form in that it prolongs life but at a great cost in terms of quality of life. Maxwell also notes that the fear of polio was also a fear of a 'hopeless existence'. From a *Life* magazine article from 1952 about life in and around polio victims living in iron lungs:

To a polio victim who can no longer breathe for himself [sic] nothing is more disheartening than the thought of spending every hour of his life on his back in an iron lung. Watching the world reflected in the mirror above him, he has one compelling desire – to escape, if only briefly, from monotonous immobility.

(*Life*, 19 October 1952, 127)

Maxwell describes the iron lung as an 'inelegant piece of technology', that physicians and nurses found awkward in terms of providing care with such limited access to the patient's body (Maxwell 1986, 22) (see Figure 11.1).

The iron lung was an airtight cylindrical chamber that enclosed – encased – all of a person's body except for the upper portion of their neck and head. An

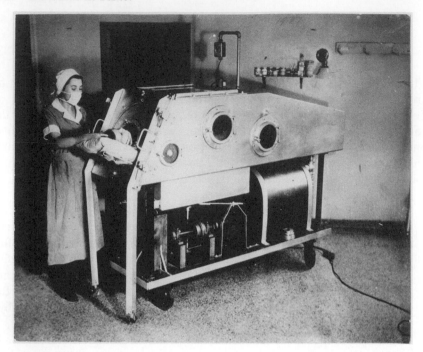

Figure 11.1 Iron lung technology

electric pump created a negative pressure vacuum to enable breathing. Early examples of an iron lung existed in the 1920s (ibid., 5–6). One radical form of the iron lung, developed by Philip Drinker and Dr James Wilson, was the creation of a room-sized respirator. One such form of the technology existed in the basement of the Children's Hospital in Boston, which could hold four or five patients, with a single doorway for staff to enter and leave through (ibid., 7–8). The room-sized respirator enabled movement of staff around and near patients but remained a situation of immobility for patients as their heads were outside of the breathing chamber, with their necks protruding through sealed portholes. Due to the large space this form of technology required, it was not widespread (ibid., 9). Major polio epidemics, such as those in Los Angeles in 1948–1949 and Copenhagen in 1952, led to the creation of specialized care units.

To this point we have considered the relational nature of mobility and immobility and provided an overview of iron lungs as a life-sustaining technology that creates an immobile situation for those living inside them. Now we turn to the accounts of those who lived in iron lungs, considering in particular the reliance mobilities that addressed their needs.

The primary source for this chapter, in terms of accounts of life inside and around an iron lung, was written by polio sufferer Mimi Rudulph (1984). Other accounts have also been investigated and considered in developing the themes that follow, while at the same time giving prominence to the first hand experiences expressed by Rudulph. Mimi Rudulph, born in 1923, grew up in Wollaston, Massachusetts. Married and a mother, she was stricken with polio while living in Los Angeles. With time, she was able to go home from her ward setting on weekends and eventually permanently, using a 'shell' type of respirator, which enabled some mobility when compared to the highly constrained existence of living inside an iron lung. In terms of the other accounts considered, Viola Pahl, lived in Alberta, Canada, was a wife and mother and was able to have a life independent of the iron lung after having spent time in one (Pahl 1954). Pahl was stricken by polio in the 1940s as an adult. Martha Mason grew up in Lattimore, North Carolina. As a girl she experienced the death of her older brother to polio and then experienced polio herself in 1948. She spent most of her life living in an iron lung in her family's home, attended by family and assistants (Mason 2003). Leonard Hawkins and Milton Lomask (1956) write about the life of Frederick Snite, of Chicago, who as a young man on a trip around the world was inflicted with polio while in China. He came from a wealthy family, and efforts were made to bring him home, where he spent the rest of his life living in a lung at family homes in Illinois, Wisconsin and Florida, with the help of family and other assistants.

From these various accounts a variety of themes are presented below. The first to be covered is entering the iron lung and the resulting transition from a mobile life to an immobile life. Second is the constraining nature of the new situation as a result of living inside an iron lung. Third is the strong desire of someone living inside an iron lung to get out. Fourth is the sense of community while living in an iron lung ward. This is followed by seeing family as mobile others. Next is highlighted the coordinated movement of people inside iron lungs, by others, in 'mobile with' configurations. This is followed by a discussion of home assistants. As a whole, these themes provide us with glimpses of the relational nature of mobility and immobility, the constraining nature of immobility and examples of intergenerational reliance.

Entering the iron lung – transition from a mobile life to an immobile life

For many who spent part of their lives within an iron lung, there were moments of shock where the formerly taken-for-granted everyday mobile life was stripped away and a new life highly constrained by the iron lung began. The person's body was being ravaged by polio, placing them in a situation where they could no longer breathe on their own. Polio was also affecting muscles that enabled walking, reaching and touching, yet without properly functioning lungs other capabilities were not possible. Here Mimi Rudulph describes her first entry into an iron lung:

> Quickly, without warning, the tall slender negro man and a white nurse lift me onto the high narrow bed they have just put sheets on, tie a white

cloth around my throat, slide the bed into a huge green metal cylinder and lock three huge clamps in front of my face.

I feel a motor running below me. My bed vibrates like a harp. Air is about me, its coolness blowing on my neck and breasts but I cannot breathe it in. I cannot find it. The choking collar around my neck keeps going in ... out, in ... out. There is air somewhere, but how can I get hold of it? How do I push it into my lungs?

Air is coming regularly. It is steady. It is slow but it keeps coming. My voice says, 'what ... do ... I ... do?' strangely. Some words come out loud, others are unheard even by me. Somehow, my life is going on. It has not stopped. The life that was swiftly leaving a few minutes ago is being stabilized by this machine with a motor under me.

(Rudulph 1984, 55–56)

In Rudulph's account a mobile life is abruptly left behind in order to sustain life. Due to polio, independent breathing has been severely challenged to the point where residing inside an iron lung is required. When independent breathing is difficult, any form of independent muscle and limb movement to bring about walking is also likely to be challenged. A new immobile life commences, surrounded by the noise of an air compressor and a reliance on mobile others to address one's needs. Throughout the accounts of living in an iron lung, there are vivid, somewhat nostalgic, remembered images of life on the move before having to enter the iron lung.

Constraining nature

Living inside a technology that encases most of the body leads many of those having lived a life inside an iron lung to express the highly constrained nature of this life. Rudulph describes not being able to scratch her nose and 'letting it aggravate until it stops itself, for nurses have no time for such unimportant minutiae' (1984, 72). A simple movement such as bringing a hand up towards one's face, often taken for granted in everyday life, is not possible while inside an iron lung.

Not only are there constraints on the person residing in the iron lung, but also on the family wanting to maintain contact with them. Here Rudulph describes the staff-imposed regime of when the family is allowed to visit the iron lung resident in her ward:

The administration feels very strongly about the length of the visiting hours on Saturday and Sunday. It is from 2–3 and on Tuesday and Thursday evenings from 7–8. '*And an hour means sixty minutes*, period,' our head nurse, Miss Hawkins reminds her lower echelons who relish kicking out lingering husbands. How short a time it is when most of them travel quite a distance. But no matter, and certainly, there can be no other time for visitors. Rules are rules. Burl [her husband] teaches those nights he explained to her, saying he would like to come Thursday afternoons for an hour. 'Absolutely not,' was her reply. So he comes anyway and she has not the gall to face him. It would be a very long time from Sunday to Saturday not to see him. Sometimes, nurses are as inhuman as doctors can be.

(Ibid., 115)

There can be distance and operational challenges that limit when and how the family makes their way to and into the iron lung ward. These can be seen as constraints on the family mobile others and also the person living in an iron lung, in that there are constraints on maintaining important family connections and social networks.

Desire to get out

Earlier on I quoted from a 1950s *Life* article expressing the desire of those living in iron lungs to get out and break free of their immobile existence. Rudulph describes the constant desire to smell fresh air and how she reminisced about New England and the smells of shrubs, trees and the ocean:

> Will I ever be able to live in the world again, to accomplish even the simplest things? Will I ever again know the sun, the air and its breezes, holding a yellow rose, touching my violin's string, all the tiniest unimportant taken-for-granted movements of ordinary existence? Will each one be denied me forever?
>
> (Ibid., 72–73)

Still later in her writing Rudulph notes:

> I want, I want, I scream, to get out of this steel house but my body and mind have no real knowledge of how to do it. Nor do they have any resources to draw on. I am drifting through each day without accomplishing anything except adding another twenty-four hours to my lifeline. Is this enough?
>
> (Ibid., 97)

The desire to get out of the lung appears strong, yet its life sustaining capabilities quickly draw the person back. From the accounts reviewed, the desire to get out of the lung seems to be strongest among those residing in a ward setting. After reading further accounts in the future it may be possible to ascertain with more certainty if this is somehow tied to having a small increase in the ability to control or influence activities in a home setting, among friends and family rather than ward staff, or other factors.

Community in an iron lung ward

Some people were able to live in home settings, with others addressing their needs. Frederick Snite, coming from a wealthy family, had assistants who catered to his needs and assisted with coordinating travel to and from events and the family's several homes. Martha Mason, coming from a less affluent background, was cared for primarily by her mother and a care worker. In both of these examples, each was typically the only person residing in an iron lung within the setting they lived in; a network of mobile others helped to address their needs. Other sufferers lived in formal iron lung wards where health care workers attended to residents' needs. Living in a ward with other people residing in iron lungs enabled residents to communicate with one another, creating a sense of community where challenges and issues could be discussed, family stories shared and issues with health care staff chewed over, often when such staff were not present. Sharing a ward meant that there was little room for any forms of intimacy. While a family was visiting their loved one, others in the ward were able to listen in and see the interactions of visits. One example:

> Terry is nine and lives two lungs to my right. Today her mother decorated a lovely white table-size Christmas tree for us. I watched her place each bauble and piece of tinsel on the perfectly shaped little spruce ...
>
> (Ibid., 76)

Here Rudulph describes the challenges of communicating with other residents, where few physical gestures are available and easy viewing of others is not possible:

> We all are in a peculiar situation because our personalities must come out through our mouths in as much as we are either in lungs or on beds, and therefore have no eye contact or body nearness with one another (as people have during conversation) because most of us cannot move our heads. Nor can we use out hands, shrug our shoulders, touch someone to make a point. We cannot shift weight from one foot to another, hold a glass or fling our heads back to get a hair out of the eye.
>
> (Ibid., 111)

In other words, many of the physical gestures and modes of action, that form part of how mobile people communicate with others, are unavailable (see examples of how physicality is part of communication in the works of Edward T. Hall, e.g. Hall 1966). In this sense those residing in an iron lung ward were constricted in their communication. Eye contact through mirrors may be possible with some residents, but not with all in a ward. No finger pointing, direct or indirect looks, shoulder shrugs, light or heavy touching can be performed or interpreted.

Family as mobile others

Already we have seen to some extent, through various accounts, how health care assistants act as mobile others, coming to immobile people residing in iron lungs in order to address their needs. Another important group, similarly implicated, is family members who are also mobile others. Mimi Rudulph's husband Burl came to visit her in the ward every afternoon (1984, 84). Here is a combined example of staff being mobile others, moving Mimi to the window, and Burl being a mobile other, bringing their daughter to the courtyard area so that Mimi can see her. Due to her physical constraints, Rudulph sees her daughter from a distance, through her ward window, where down below her husband and daughter have made the journey to be close to her:

> I live all week long for Sunday afternoons when the nurses push my lung to the window where through the closed glass I watch little Deri scamper around down in the courtyard on the blanket Burl brings to put her toys on. A year old, December 22, she runs everywhere and plays with Big Carol, Doodie and Howard. It is fun being with her, even from a distance. Her little blonde, bubbling, bursting, buoyant self has no idea how much life she transfers to these hungry eyes so thrilled and happy to see her. She is life to me ... puckish, full of fun and humour, the being I love more than anyone I have ever known.
>
> (Ibid., 87)

In the early stages of her life in an iron lung Martha Mason describes visits by her father:

> Every weekend my father brought all sorts of things people from home sent me. Because my picture had been in the Morganton newspaper, strangers dropped by the hospital with books, flowers, candy, and other gifts.
>
> (Mason 2003, 179)

From these accounts we can see family as mobile others coming to their loved ones who reside in iron lungs. Indirectly we also see staff performing a 'mobile with' as they wheel Mimi to the window. We also see others from the broader community as mobile others, bringing books and other things to Martha.

Coordinated movement by others ('mobile with')

So far we have seen a variety of mobile others helping to address the needs of someone who is immobile. There can also be 'mobile with' (Jensen 2010; 2013) configurations that help to address the needs of someone who is immobile, living inside an iron lung. Here Mimi Rudulph is part of a 'mobile with' configuration, within which she has limited capabilities to contribute as part of the arrangement:

> Through my cracked isenglass mirror I see the police escort pull in ahead of us as we reach the hospital exit. Hearing sirens pierce the noon hour's

city business, I know there is no turning back. It is my turn, now, to live again. I am coming world, ready to begin again.

On and off the freeway we speed, my body swaying with the motion of the ambulance's weaving through detours of freeways being re-constructed. At main intersections cars slow or stop as we speed by, freezing their occupants as they see an iron lung with its bellows moving up-and-down, up and-down, keeping alive the head they see sticking out of it. I see fingers point my way as heads turn away. Little do they realize the happiness of 'that poor soul.'

(Rudulph 1984, 105)

From these accounts we can see a strong reliance on others and technologies to bring about 'mobile with' movements. In each of these examples, the person inside the iron lung has little to contribute to the physical aspects of the 'mobile with' arrangement.

Residing in an iron lung, with little or no ability to perform any independent form of mobility, and little to contribute to 'mobile withs', there is a strong reliance on others to address needs. Now we can turn to requesting that a need be addressed versus the actual addressing of that need. Here Martha Mason describes seeking to be turned at a particular point in time, while ward staff had a different point of view as to when a turn should take place:

Only one nurse condemned our activities. This cocky, strutting woman had already caused those of us who couldn't turn ourselves a great deal of physical pain. She insisted that we turn at night by the dictates of a clock instead of our needs. If I, or any other immobile person, awoke in pain from pressure points, we were forced to wait until her clock's hands showed mercy on our anguish. Neither our sleeping patterns nor our bodies were alike. Thin hips and shoulders hurt before those better padded with fat. Youngsters whose sleep was invaded by nightmares could often escape the marauder by changing positions. This 'Nurse from Hades,' as Mrs. Lee dubbed her, was in charge of the house. Miss H's word was law.

Often she chased my portable friends out of the 'respirator room' and back to the 'big room' across the hall. On one occasion, she shook her finger at me and said, 'Be friends with these girls. They're like you. You don't need to be fooling with people going home. You'll be here with Kay and Lena for as long ...'

(Mason 2003, 196)

At times such as in the account above, there are glimpses into the inter-generational nature of the dynamics at play, where young and middle-aged nurses and doctors are caring for children and adults living in iron lungs. In terms of a housebound senior, there may be instances where a senior is keen to 'get out' or visit a friend, and who then has to negotiate mobility assistance with younger friends, grown children, grandchildren, or others, in an attempt to fulfil their desire for mobility.

Home assistant

For those living in iron lungs outside of a formal ward setting, a network of support was established in home settings. Just as we have seen in a ward setting, here too, having others being mobile on one's behalf, attempting to address the needs of the immobile, does not mean that all goes as the immobile person would like:

> On her frightful days, I had neither bath nor other care except for bedpans – perhaps only three times in an entire day. I sent Ginger to a restaurant for food. Many days, I ate nothing – not because we had nothing to eat but because I was full of despair. There was no room for anything else.
>
> (Mason 2003, 45)

And the role of assistants can change over time. Here Martha Mason describes how more and more duties of care were being transferred from her mother to her assistant, Ginger:

> More and more of my personal care was delegated to Ginger. Together they bathed me and changed my bed. Both of them turned me except at suppertime, when Mother and I were alone. She was able to remove a pillow from my back and two others from between my legs and then drag me into a relatively straight and comfortable position.
>
> Closing the iron lung was difficult for her because of her arthritis and her osteoporosis. My guilt no doubt added greatly to the weight of my sliding bed, perhaps because the idea of her not closing the clamps had occurred to me. The vise-like hardware must be secured in order to create a seal so that trapped air under pressure can force air into and out of my lungs. Unless there is a tight seal between the cot carriage, which rolls out of the shell, and the cylindrical tank, the machine is useless. I don't breathe!
>
> (Ibid., 49–50)

Conclusion

From the accounts of those who lived in an iron lung, we can see the relational nature of mobility and immobility. Those who are immobile are, to varying degrees, relying on the mobility of others in 'mobile other' and 'mobile with' configurations, in order to address their needs. In one sense, this teaming up can be seen as enhancing motility (Kaufmann 2002). Independent mobility for those who resided within iron lungs was not easy. In the absence of independent mobility, needs were not always addressed where and when an immobile person would have preferred. Different levels of negotiating were required and, at times, compromises were made by the immobile and mobile in order to address needs. From these examples, the immobile person residing in an iron lung does not always passively accept how their needs are addressed. In a preliminary sense there seems to be a stronger desire to 'get out' of ward settings to a home setting. Future research may be able to establish whether this is somehow related to having a stronger role in deciding how needs are to be addressed. In many cases there were vivid recollections of a past mobile life that had become no longer possible. Infrequent journeys, where a person residing in an iron lung is moved through configurations of mobility, appeared exhilarating for the immobile person.

We can now turn to what lessons there may be from this investigation, with regards to those researching and planning housing, health care, social services and transportation systems for housebound seniors. In both cases, among those who resided within iron lungs and seniors who are housebound, there is a diminished capability to perform independent mobility. For those living in an iron lung, muscles were ravaged by polio, making independent breathing difficult. For a housebound senior, there may be problems with walking and balance that make movement within their own home difficult, and outdoor movement even more of a challenge. If they used to drive a car, it is likely that this ability slipped away before they became housebound. In both cases, earlier life may have included high degrees of independent mobility with the car having been a resource within 'mobile with' and 'mobile other' configurations, to assist with the needs of others. Both the iron lung person and the housebound senior are highly reliant on the mobility of others to address their needs, and in many cases family members performed such tasks. This harkens back to Martin-Matthews (2007) noting the importance of considering the 'other players'.

Similar to some iron lung people living in a communal setting, with formal care workers, while others reside in their traditional homes with a network of family and friends providing support, one can also see this among immobile seniors. Some move to a setting where there is formal care and support workers, such as long-term care homes and retirement residences. Those who remain in their traditional homes may seek out formal and informal home care and family assistance to address their needs. Both scenarios alter the related constellations of mobility, be they family, friends or others. In a communal setting, a group of immobiles are congregated and others travel to and from this location addressing the needs of the immobiles while in that location. In a traditional home setting, the mobility constellations are typically geared to addressing the needs of one immobile person. Seen at a more macro level, efforts need to be drawn to the spatial elements of attempting to serve the needs of a dispersed group of immobiles. Those planning for the needs of housebound seniors need to pay close attention to where older, potentially soon housebound, seniors are located and their ability to call up resources to address their needs. This could involve modelling for delivery at a regional and neighbourhood level, and needs to account for all aspects of ensuring quality of life.

This chapter has opened an investigation into the reliance mobilities of those who lived immobile lives, within iron lungs, and those who interacted with them, such as family, friends and health care workers. This inquiry may be of assistance in looking at how to address the needs of housebound seniors.

12 The global urban space
Older age and Filipinos as global workers

Karel Joyce Kalaw

Introduction

This chapter aims to offer an understanding of the intersection of ageing, life course and mobility studies. Here mobility refers to the movement of people from one place to another in a global sense and the ways in which it impacts the everyday lives of overseas contract workers. The structural force of demography, for example labour migration as a driving force and strategy for a good life for one's family, is key in understanding the inclusion of migration as a life trajectory in an individual's life course. This study focuses on the public event of labour migration that impacts the personal and intimate lives of a migrant worker's life. The case of the Overseas Filipino Worker (OFW) returnees specifically illustrates the intersection of ageing, life course and mobility studies. The chapter highlights the role of culture in understanding this intersection.

I am drawing from Massey's (2005) idea of space, such that spatiality is constructed from the exchanges between the global and the most intimate dimensions of human life. In Massey's terms, space is not solely confined to being an absolute element, but is socially produced and is always renegotiated. With the idea of space, the notion of trajectory is integrated such that several, different trajectories simultaneously and interdependently exist (Jirón 2014). The movement from one place to another and being in place suggest that changes in an individual's life course may occur over time.

With this in view, this chapter seeks to extend the notion of global urban spaces to the intersection of self and society as refracted by age. The Philippine case, in which labour migration is particularly relevant, opens up an opportunity for discourse across fields of knowledge. Hence, the study has three specific aims. The first is to invite a broader understanding of mobility studies exemplified through labour migration as it intersects with ageing by using the life course perspective. In so doing, this chapter hopes to illuminate an understudied and less explored area and go beyond the economic rationale of migration while investigating the inner subjectivities and the socio-psychological dimensions of return migration in later life. The study extends the few studies on labour migration addressing the simplistic, two-fold divide of micro- and macro-explanations (Porio 1999). The experience of labour migration bridges this disconnect, as the event of return migration cannot be easily factored into this simplistic binary gap, but must be understood within the dynamic interrelations of both structure and agency.

The second aim is to extend C. W. Mill's discourse on linking the personal experience of old age with the public affair of labour migration. Mills (1959, 12) once stated that, 'No social study that does not come back to the problems of biography, of history, and of their intersections within society, has completed its intellectual journey.' The experience of old age intersects and pervades personal and public spaces. A good example to illustrate this point is the case of labour migrant workers returning to their homelands, their eventual later life experience and how their return experience falls into place with the entire rationale of doing overseas work. In this chapter, I explore the possibility of taking a life course perspective in understanding the lives of migrant workers as they experience the life events of return and older age. I suggest a conceptual map to make sense of the nexus of migration studies and later life. The public understanding of labour migration is brought into light with the life course framework as it takes into account the personal event and significance of older age among returning migrants.

Third, I seek to extend the use of indigenous methodologies (IM) as a culturally sensitive approach to doing research. Indigenous methodologies highlight the different spaces, for example physical, socio-cultural and psychological, that would allow us to understand the intersection between migration and ageing; and describe the inner subjectivities and humanness of migrant workers as they return to their home country. Filipino research methodologies,[1] developed from the inspiration of IM, enhance traditional qualitative modes of doing research through the application of culturally-fitting data collection techniques, for example *pagtatanong-tanong* and *pakiki-pagkwentuhan*, which will be discussed later in the chapter.

The focus on Philippine labour migration offers a unique opportunity to investigate the potential exchange between the life course perspective and migration studies. The eventual progress of migrant workers to older age warrants understanding of how they see their ageing experience. The case of the OFW, their very location in the flow of global migration, places them in a good position to illustrate Mills' statement; such a personal affair as ageing is juxtaposed with the public event of labour migration. The chapter provides a discourse on less explored dimensions of migration. The first concerns the inner subjectivities and humanness in migration studies as much as migration scholars are more drawn to investigate the macro factors, for example in economic, political and historical accounts. The second dimension highlighted is the return experience of male OFWs as being a 'silenced' population in line with the feminization of migration. This is so since less attention is given to men who are unintentionally marginalized. The chapter offers a male story of return experience and suggests a holistic, gender- and age-sensitive understanding of migration (i.e. return migration). Hence, the significance of the study lies in its contribution to the literature, to scholarly pursuits, and to the potential formulation of policies over time.

Context: Philippine labour migration and the overseas Filipino workers (OFWs)

> Migrant workers are an asset to every country where they bring their labour. Let us give them the dignity they deserve as human beings and the respect they deserve as workers.
>
> (Juan Somavia, director general, International Labour Organization (ILO), 2010)

The common trend of developed nations attracting workers from developing nations to pursue international labour migration draws several academic and policy interests, each offering their multiple and varied perspectives. As evidenced in Asia alone, which is home to 58 per cent of the world's population (Hugo 2004; Asis and Piper 2008), the pervasiveness of labour migration ranges from unskilled and low-skilled to high-skilled workers, and often it involves temporary contract labour (Hugo 2004). The extensiveness of such migration on the international and global scene necessitates an investigation and understanding of this phenomenon.

The Asian labour migration scene is characterized by: (1) having a temporary migration agreement with the host countries; (2) being both interregional and intraregional; (3) having a large number of undocumented migrants; (4) having a well-developed migration industry; and (5) having a relatively large number of female migrant workers (Asis and Piper 2008). This trend will continue to persist as long as there is demand from developed nations for the surplus labour provided by developing nations. Moreover, the landscape of international Asian labour migration is changing, as more highly skilled professionals are needed by developed nations to address the needs of IT and healthcare, yet the demand for the '3D' (dirty, dangerous and difficult) jobs (Massey et al. 1993), or those of low-skilled workers, cannot be overlooked and will always remain.

In understanding the Asian manifestation of the globalized labour market, it is important to highlight the role of the Philippines as one of the world's leading labour exporting nations. The countrywide statistics are informative: an estimated 9 million overseas contract workers went abroad to work in 2006 and sent remittances to families and relatives in the Philippines amounting to US$12.8 billion (Porio 1999). These numbers represent more than 10 per cent of the Philippine population and about one-fourth of its labour force (Porio 1999). Filipinos can be found in almost all countries and territories of the globe (Asis and Piper 2008). Men and women are equally likely to emigrate, but in more recent years there has been a feminization of labour. Overseas Filipino Workers (OFWs) range from less skilled to highly skilled workers, but more important is sensitivity to the labour market. The Philippines is more likely to supply less-skilled workers in the international labour scene. Together with the Philippines, Indonesia, Bangladesh, India, Pakistan, Sri Lanka and Nepal (Asis and Piper 2008) all share this feature of being labour-exporting nations.

OFWs are commonly referred to as the 'modern day heroes of the nation' primarily for the remittances they contribute to their home country's economy, growth and development. The export of OFWs and their sizeable contribution to the Philippine economy in remittances is consistent with global trends, in which remittance flows to developing countries rose from US$31 billion in 1990 to US$167 billion in 2005, with the Philippines ranking fifth (after India, Mexico, China and Pakistan) in the total value of remittances received from abroad (World Bank 2006). The vital contribution of remittances to local economies has been particularly illustrated in the last ten years, when remittance growth outdid private capital flows and official development assistance (ODA), for example in 2004, when remittances were larger than both public and private capital flows (Porio 1999). Furthermore, remittances have tremendous effects on the organization and reconstitution of household economies in the places of origin and places of destination over time. These changes are especially significant as migration and remittances have been a central strategy for many Filipino households over the last forty years or so. Migration has traditionally served as a means of alleviating the difficult lives of individuals and families, and hence becoming an OFW sustains not only the nation's development, but also improves family economic situations.

The life course perspective

> Labour migration has become a part of the calculus of choice for people throughout the region.
>
> (Hugo 2004: 77)

There are a number of perspectives on return migration, from more traditional views to multi-disciplinary ones (Cassarino 2004; Cerase 1967; Massey et al. 1993; Asis 2006). With these multiple perspectives in mind, it becomes apparent that the rationale for the return is both economic and non-economic in nature with issues such as family links to the homeland being of consequence, and broader socio-political issues in the host country (Stinner et al. 1982). However, motives for return cannot be summed up neatly, as return migration is a multifaceted social reality that is contextualized and experienced differently. Given this complex set of factors in play around return migration, the life course perspective becomes a useful framework in making sense of the return experience of OFW and family members. The following will discuss the assumptions of the life course framework, and then incorporate its utility in the discourse of labour migration as a whole. The life course perspective is then specifically applied to understanding the return phase of migration as experienced by the overseas migrant worker in later life. First, it is necessary to explicate this perspective in relation to the life trajectory of migrant workers. This is evident in Giele and Elder Jr's (1988) four key elements that, they argue, influence an individual's life course: human agency, linked lives, historical and geographical context, and timing of life events. Within this framework human agency allows the person to make conscious and voluntary decisions in order to create one's life path. The life course perspective permits contingencies in trajectories and discontinuities in transitions. The concept of linked lives highlights the individual's embeddedness in social relationships and with other people, as well as one's awareness of the saliency of other people in the formation of one's transitions and trajectories. The historical and geographical context refers to the 'lives in time and space', or the cohort effect, which affects the life paths of individuals. The timing of lives includes the occurrence, duration, and sequence of transitions (Elder Jr 1994). The emphasis on the timing of events in context of the several life domains of a person embedded in social relations makes the life course appropriate for viewing the intersection of life course, later life and labour migration.

The life course framework as a theoretical and conceptual tool needs to be mindful of and responsive to today's changes in society. It should enable an understanding of the emerging social realities prevalent in society, linking the four premises of the earlier understandings of the life course elements of human agency, linked lives, historical and geographical context, and timing of life events. Dannefer (2003) acknowledges the need for a life course framework responsive and sensitive to societal changes and emerging patterns of social life, and has presented a global understanding of the utility of the life course framework. This global understanding is applied to his discussion on the conceptual distinction between life course as a phenomenon and as an explanatory theory. Within the life course as phenomenon approach, the changes emerging and occurring in the social and family set-up, the institutions of the life course are reworked – 'the "typical" life course is not at all typical for much of the human population' (Dannefer 2003). Dannefer's study proves useful in investigating the link between labour migration as it impacts life chances, the structure of opportunity and the meanings of the standardized 'three boxes of life' and 'personal life'. With the life course as an explanatory theory, early life events and experiences are seen to impact the direction of a person's later life (Dannefer 2003). This highlights the predictability of life routines and the capacity to map life in ways based on societal features and its general rules of conduct as well as maintaining societal balance (Dannefer 2003). This offers a useful opportunity to present the event of labour migration, and particularly return migration, as a disruption of the general social orderliness of OFW returnees' life course and its eventual impacts on their family. Life course perspectives invite several levels of analyses when looking at labour migration. These life course principles do not comprise a comprehensive life course explanation, yet are informative in describing the highly complex and dynamic interconnectedness of structures and personal accounts over time.

Life course and the return experience

The discussion above helps in the reconsideration of migration models and takes into account the concepts presented by the life course perspective. I am using the terms 'self' and 'individual' interchangeably to refer to the OFWs' state or quality of being an individual or person separate and distinct from other persons and with their own needs and goals. As the discussion that follows illustrates, the notion of familism as a Filipino value is a salient reason for overseas work. The desire to provide a good life for the family is an overseas worker's mission – to earn, save, and go back to the Philippines after the completion of the migration 'project'. Migration is spurred and translated into reality because of familism. The trajectory of return of the male OFW is included into the greater nexus of trajectory systems in an overseas migrant worker's life course. Return migration and the life domains (with the inclusion of migration) interact with one another and are responsive to each other, such that the concept of linked lives, the role of family and other networks in fostering the migration move, and the embeddedness in the societal structures as defined by institutions, are noted. I suggest that the relationship between the trajectories of family, education and employment is not unilinear; rather it is circular, representing the dynamic fluidity of its interconnectedness. The return is the seeming endpoint of the 'journey', yet this is not the case. The return prompts and reconfigures the already restructured family dynamics, skill acquisition, and prospective consideration for employment, for example self-employment as an outcome of the actual migration itself. Rather than using the temporal, normative and structural ordering of education, work, and retirement in the conceptual map, I refrain from the use of retirement as a construct, as there is a likelihood that OFWs would not see their return as retirement. In the final analysis, whether the migration project was successful or not upon return, the OFW, in his later life, needs to contend with the objective changes and realities he now faces. By incorporating the life course lens, we can see the interconnectedness among parallel trajectories of labour migration (return), family, education and employment (and potential emerging life trajectories) as situated in the migration process.

Research methodology

The research, informed by ethnographic interviewing, life stories, narratives and Filipino Indigenous Methodologies (FIM), was conducted in a small town in the Philippines from June 2013 to January 2014. Ethnographic interviewing allowed me to establish rapport and respect sufficient for a genuine 'meeting of minds' and develop a collective exploration of the meanings the researcher applies to their social world (Heyl 2001). The stories people tell about themselves and their lives generated by the life story approach (Linde 1993), allowed me to make sense of people's experiences. The use of narrative method is informed by a social constructivist perspective (de Medeiros 2014) that allowed a human way of making sense of the stories shared and offers meaningful interpretations of the world through temporality (Polkinghorne 1998; Smythe and Murray 2000). The FIM[2] included *pagtanong-tanong* (*Tagalog* word meaning asking questions) and *pakikipagwentuhan* (*Tagalog* word meaning telling and sharing stories). *Pagtanong-tanong* allowed the casualness and friendliness of asking questions (*pagtanong-tanong*), which led to the participants being more comfortable in telling their stories. Similarly, *pakikipagkwentuhan* evidenced the relational nature of Filipinos with the informal and friendly tone it exuded that encouraged a trusted, unobtrusive, and safe venue for participants to share their stories. These methods allowed me to elicit rich data from the participants' stories. Two samples were interviewed: the OFW returnees and their families. A total of six OFWs, four OFW wives, and five OFW children were interviewed. For comparative purposes, I also interviewed non-OFWs, 64 years old or older who had remained in the Philippines and worked and lived for a significant amount of time in the small community. A total of four non-OFW, three non-OFW wives, and four non-OFW children were interviewed.

Reliability was established in the following ways: analysis was peer-reviewed by colleagues familiar with qualitative analysis; repeated interviews were done to ensure richness of data; the stories were member-checked to add rigor to the analytic process (DePoy and Gitlin 2015); quotes were included and translated to English to enhance and substantiate stories, creating 'thick descriptions' (Geertz 1973); a comparison group was included; trust and rapport – the essence of *kapwa*[3] – was built with participants in order to elicit more comprehensive and truthful data; biases and limitations were recognized; and notes on reflective observations, thoughts and descriptions during the course of the data collection were made, recorded, and referred to when analysing data. These included transcripts, field notes and memos, and conversations with professors and colleagues. Finally, I used a cross-participant comparison grid based upon my research questions, emergent themes and special findings to sort and organize the data, and carried out a thematic analysis. These themes help explain the linked lives and networked characteristics of OFWs.

Labour migration and the life course

The discussion above aids in a reconsideration of migration models to take into account the concepts presented by the life course perspective. The concept of linked lives and the role of family and other networks in fostering the migration move, together with understanding migrants' behaviours – with their corresponding cumulative effects, which are embedded in societal structures and defined by institutions – make the life course perspective appropriate for an increased understanding of migration. The case of the OFWs illustrates this point, captured by two key themes on familism and trajectory.

Of course, your utmost priority is your family. People live for their families.
(Tirso)

Yes, this is what I am thinking. I will do all sacrifices for the welfare of the children. It is all for the family. I don't mind experiencing pain and agony as long as it is for the family. I can do all for my family.

(Ben)

I am always thinking about the welfare of my family. It is supposed to be like that. It is my aspiration to provide a good life for them.

(Miguel)

The research data showed that familism, a Filipino value, prompts OFWs to engage in overseas work. Filipinos take pride in their familism. The desire to provide a good life for the family is an OFW's mission: to earn, save, and go back to the Philippines after completion of the 'project'. With the event of the return the male OFW has now to address the reality that he has grown old through the overseas experience. Miguel articulated this: 'I get tired as I am getting old. Unlike before when I was younger.' This was also shared by Tirso as he continued to drive and troubleshoot his own jeepney.[4] His strength was often tested as he maintained the jeepney. Tirso remarked:

Another thing is that when you are older, there are many changes. There was a time when my strength was tested whenever I work with my jeepney. I can't work that long and I can't move around as I used to. When I sit on the ground for a long time, then I have a hard time getting up. My knees hurt. It is not that easy as when I was younger since I was more flexible. It is the case that I am weak now. I am not as strong as I used to be since I don't have any sickness at that time. I was stronger when I was younger. This is my life now. This is the thing that I observed.

Regardless of the success of the 'project,' the older OFW has to contend with the objective physiological changes and realities he now faces, which are at the same time interconnected with the parallel trajectories of migration, family, education and employment as situated in the migration process. In the study, labour migration was seen as a key point in an OFW's biography and as an interruption to the institutionalized life course framework of the 'three boxes of life' (Riley and Riley 1994), and the 'personal life' (Hagestad 1998; Uhlenberg 1978; Uhlenberg and Mueller 2003) that impacted the life course construct over time. Miguel shared, 'Yes, my life as an overseas worker sustained my family. My being an OFW is part of who I am. I am thankful for the things I experienced abroad.' Further, Rene expressed the same thoughts on being an OFW:

The idea that I am an OFW like the rest of the OFWs means that I have something to say about my life. I sent my children to school and this is the best aspiration a father can have. If you are an OFW and you have proved yourself to provide a good life for your family, this gives us fathers satisfaction and contentment in our lives.

Both Miguel and Rene understood their status as OFWs. The interruption of working overseas offered a rethinking of the life course perspective in relation to how labour migration and culture affected individual experience and impacted the decision to migrate. The individual's cognition of their own selves, e.g. their individual trajectory, intersects the collective understandings with their family. In the study, Miguel acknowledged that he still wanted things to happen for him but took another perspective.

> I have aspirations, but it is not for me. My aspirations remain to be for my family. My aspirations now are for my grandchildren and to see them successful in life. I am here for them and make sure that they become successful and have good lives. This is my aspiration. As for me, I am fine. I have no more aspirations for myself.

Miguel's aspirations were not for himself, but directed towards his family. This was also true for Boyet as he aspired for good health for him to be instrumental in helping others and particularly his grandchildren:

> Our happiness is for us to be healthy and to be strong. We pray that God grant us strength so that we may care for our grandchildren. We don't aspire for more money, but if it comes, then we welcome this [laughs]. But to aspire for more is not our main concern. We have this life now that we are able to eat daily and this is good. We ask the Lord to grant us more blessings so we can share it with others, our relatives and our siblings. This is my hope, to help others whenever we can. The life we have now is good, and we are content with what we have.

Miguel and Boyet's personal aspirations demonstrated that familism was a compelling force to act and do things for their family. Though articulated as something personal, the end product of their aspirations was for others and for their family. These accounts proved that the rationale for doing overseas work, as exemplified by familism, continued to be a potent force upon their return. The value of familism kept the family sustainable and together over time, such that familism was nurtured even more in their later lives. Familism is thus a potent value in the lives of the OFW returnees.

The study generated a new understanding of the premise of the life course perspective as it relates to the consequences involved in transitions, which 'are discrete and bounded; when they happen, an old phase of life ends and a new phase begins' (Hutchinson 2005). For Greg, grandparenthood opened up an opportunity to take on his grandparental role. 'I take care and attend to the needs of my granddaughter now. I bring her to school and pick her up as well. I make sure that I make myself busy especially now.' For the male OFW returnees, the return meant the recognition of their aging. Going home for good provided an understanding of the old, emerging and new roles they have at home with their families.

The life course perspective as an individual and biographical affair was brought into question as the saliency of linked lives and interdependence of individuals. These centre on the family as the basis for synchronizing the individual's life with family members' lives over time (Oleinikova 2013). Tirso explained that the decision to go home was collectively agreed by his family:

I don't have any aspirations or dreams. I think about and prepare for my return. Just in case I wanted to work overseas again at that time, I can. But I decided that since I am getting old and the plea from my family, I decided to finally go home. Now I am driving and at the same time managing our farmland. And this is I think is good enough.

His return allowed him to think about his ageing. He had imagined returning when he was younger, but realised that he would go on working overseas into his later years. Tirso saw his return as a recognition of his ageing. Though he gave up overseas work voluntarily, he envisioned his return as an acknowledgement of his ageing. He was set into his later life as he saw himself doing the things he did prior to overseas work. It is important to highlight that familism, again, is the value that drove the entire migration process. As such, overseas work and the eventual return was a collective strategy and commitment among family members. It can be said that this familism emanated from *kapwa* as an ingrained and core value among Filipinos. *Kapwa*, being the shared otherness and high regard for the well-being of others more than oneself, validated the claim on familism by the returnee's family.

The role of culture is embedded in social, economic, political and historical arrangements and is shown to guide the returnees' life. This begins when labour migration maps the life course of individuals who decide to work overseas to provide a better life for their family. It could be said that labour migration is a collective understanding and action to counter the social degradation (Okólski 2004) of family members. As shown in the study, the familiar and standard understanding of the life course perspective of the high regard for human agency in making choices (Hutchison 2005) and the use of personal power to achieve one's aims (Oleinikova 2013) were rethought. To extend the usefulness of the life course perspective, it is argued that human agency should not be privileged and individual differences should give precedence to collectivities and social groups (Elder 1994; O'Rand 1996). The individual event of migration, in this case return migration, is collectively understood and executed by family members. In the study, being a Filipino has been never an individual event and the individual and personal events are mostly collective ones. A Filipino individual is made up of a village within; it breathes the shared otherness – *kapwa* – among the male OFW returnees.

However, it is also important to record the contradiction that familism espoused. As Gardener (2002: 226) argues, 'migration is inherently contradictory for it involves physical separation in a society which so greatly values togetherness'. Familism is redefined and given new meanings in contrast to the notion of togetherness (Medina 2001). Tirso describes his experience of being absent from his family: 'Homesickness will always be there but you have to fight it. My thoughts are refocused in giving a better life for my children and have them placed in good schools. I can't stop and be defeated by nostalgia.' The absence of a family member in the person of the OFW working abroad accomplishes an important wish of the family to have a better life that impacts family structure, formation and relations, and household resources (Bretell 1988; Lagrosa 1986). With the prolonged separation of the OFW from his family, feelings of guilt and a sense of disorientation impact both the returnee and his family (Sri 2009). The study showed that even values are responsive to the emerging structural changes prevalent in today's society.

The return experience and the entire labour migration process is a social phenomenon responsive to the structural changes, which following Dannefer's call for studies to enrich the field of life course studies, need a more careful inclusion in this approach. This is in response to a plea to enrich the field of life course studies and to widen the applicability and scope of its existing knowledge base (Dannefer 2003).

In sum, the case of the OFW returnees showed that individual choices are embedded in the family and are delimited by Philippine economic, political, social and historical arrangements as well as global interdependencies. The life course perspective pays attention to the effects of historical and social change on human behaviour (Oleinikova 2013) that are significantly relevant to Philippine society. In this case, labour migration, in this case return migration as experienced by the OFW returnees, was a collective rational decision in light of the assessment of one's family's welfare and future aligned with its cultural values.

Conclusion

At the beginning of the chapter, the notion of mobility is extended to a global sense and then translated to labour migration as experienced and told by the male OFWs upon their return. From their stories, the return experience is recognized as part and parcel of their recognition of old age. The use of FIM enhanced the traditional qualitative methodology and proved useful in high-lighting the cultural nuances by which labour migration was brought forth. The public affair of labour migration was linked to the personal and house-hold dynamics of the male OFWs and their families and extended to the field of mobility studies.

Migration remains pervasive in human life, affecting the most intimate affairs of men and women. The case of the male OFWs shows that the experi-ence of migration reconstructs the migrant simultaneously with the economic, political, cultural, and social milieus he inhabits. Migration, as a process, disrupts the lives of the male OFWs as well as those of the significant others who surround them. Lives are changed, interrupted, and discontinued vis-à-vis one's normative life path. Migration, then, is instituted, incorporated, and transformed to become part of a person's domain of life. Ageing, on the other hand, is inherent in human life. All are bound to age, and that in itself is a plausible justification for the urgent need to integrate it across disciplines and the saliency it bears to persons situated in their specific milieus. Migration and ageing run parallel to one another in an older OFW's life and also incorporate the important role of culture. It is for this reason that use of the life course perspective is ideal for situating the older OFW's life path in relation to other domains in his life.

Migration studies share the same project that gerontology expresses: the quest for an integrated and unified theory. The complexity of migration as both a process and a behaviour cannot be neatly summed up in such a structure (Castles 2010). But important are studies that centre on the complexity, contradictions and uncertainties of outcomes in migration. That these features are seen as counter-logical are to be celebrated and embraced, as the object of study is highly complex. Opting for a middle-range theory, rather than a unified theory of migration, does not imply mediocrity in doing research, but summons a sensitive and grounded endeavour in understanding migration. In doing so, science is not compromised, as analysis holds true in identifying patterns and heterogeneity in specific types of migration. It is not the aim of a middle-range theory to predict and generalize migration in every aspect, but it invites other disciplines to engage in migration conversations. This chapter hopes to generate these conversations, to elicit dialogues with the life course perspective and gerontology. By using the life course perspective to showcase the labour migration experience of OFWs, this chapter indicates its potential as an analytical lens to explore and exploit labour migration, life course, and gerontology all together. Embedding the life course perspective within migration and mobility studies enhances the analytic power of each, and in fact reciprocally enriches each element. In the process of this research, care has also been taken to integrate culturally sensitive and appropriate methodologies necessary to draw data that respond to the aims of the study. The use of an endogenous methodology – FIM – was illustrated in the study. Thus, these research strands, when taken as a whole, can inform and instruct each other to nurture a better understanding of migration, life course and ageing.

Acknowledgements

I would like to thank Dr Kate de Medeiros for providing inspiring and constructive insights on an earlier draft of this chapter. Appreciation is also extended to Dr Carl Dahlman for the valuable expertise he provided during occasional consultations.

Notes

1 Beginning the 1970s, Enriquez (1975; 1985), together with other Filipino social scientists, called for the indigenization of a suitable and relevant social science

which would be sensitive and responsive to Filipinos and Philippine society (Enriquez 1985; 1975). They developed indigenous methodologies to elicit and represent social realities with a cultural lens that goes beyond the surface to unearth deeper meanings of Filipino behaviour, thoughts, values and experiences. Through the use of Filipino Indigenous Methodologies (FIM), my study builds on the creation of knowledge which will inform culturally sound policies over time. This is a step closer to answering the plea for an authentic and credible picture of the persistence of migration in Philippine society and the event of old age among Filipinos.

2 An indigenous methodology.
3 A Filipino word that translates as a deeper sense of shared and collective understanding of otherness; a core Filipino value.
4 Mini-shuttle bus used for public transportation.

13 Conclusion

Susan Robertson

Our contributions to the burgeoning dialogues on mobilities have explored ways that help us to establish the significance of relationality, age and lifecourse. Asking questions of what is meant by generation and intergeneration, as applied to mobilities, has opened up different ways of examining interdependencies that position the work as moving on from careful analysis and into the realm of the propositional, exploring different ways that perceptions and conceptions of the lifecourse may be understood. The authors of the chapters included in this book take diverse approaches to these considerations and together provide a rich and multilayered understanding of the impacts of intergenerational mobilities, in terms of perception, anticipation and the materialities of engaging in the desires and limitations that arise. Our contention is that generations shape mobilities and mobilities shape generations and that neither of these shapings are even, steady or necessarily one-way. It has become clear that to look at age as a generational definition is not adequate; generation as a concept is fluid, relational, intersectional and co-productive of dynamic interdependencies.

First we examine how authors have questioned the notion of 'generation' as a distinct stage of the lifecourse, which becomes apparent through mobilities. In Chapter 7, Dubucs, Pfirsch and Schmoll study the experiences of recent Italian immigrants, for whom 'generation' has become a label that identifies a group of those in early adulthood who have left their country of birth for better opportunities, and recognition of their value, elsewhere. Their understanding of this stage of lifecourse relates predominantly to their educational achievements (most being highly educated) and their engagement with employment that they judge as appropriate to their qualifications and aptitude. In this analysis it would seem that there is a mismatch between the expectations and frustrations of those seeking to enter into another generation, not specifically defined by age but also defined differently by those identified in this or an adjacent generation. Ricci, in Chapter 10, discusses the mobilities of a generation of young people who live on an estate in Bristol who feel labelled and disadvantaged by not being old enough to claim benefits but still having to pay adult bus fares. Stratford, in Chapter 2, describes the anticipations of those in one stage of life and how they are expected to behave and the responsibilities that are perceived as defining a phase of life such as adult-hood, suggesting that stages of the lifecourse are sought after and can only be achieved successfully within specific parameters of how to flourish within each stage. For each of these authors it seems that the thresholds of genera-tions and stages of the lifecourse are not fixed, and it is not unusual to experience different stages simultaneously, by remembering, anticipating and acknowledging how others perceive a generational identity.

Second, and related to the fluidity of generation, is the tendency to associate particular mobile activities with particular generations or lifecourse stages. In Chapter 6 Harley critically examines mobile phone use according to bands of ages that are more or less accepted in social science as suggestive of stages of the lifecourse. However, Harley challenges these normative assumptions, indicating that attitudes to communication technologies cannot be simply categorised according to age. Similarly, Sayago et al. show, in Chapter 5, that there is a willingness, a desire and even a necessity for older people to engage in techno-logies, more commonly considered to be the preserve of much younger people. These are required in order not only to enable continued communication with multi-generational families but also to provide the opportunity to commu-nicate with others, which is relished as a way of opening up broader social networks.

Third, the collection helps explicate the relationality of generation, mobilities and immobilities. The question of generation in Fisker's Chapter 11 arises in a 'mobile with' situation that allows Mimi, in an iron lung, to be moved to the window to see her young child playing in the courtyard outside the hospital ward. Fisker makes the case that there are useful lessons to be learned from the potential of mobilities that are enabled by family, or others, that may inform future planning for less mobile older people living in situations where the interaction with family and others is altered. Both Klinger (Chapter 9) and Gilroy et al. (Chapter 3) are concerned with the circular affects of transitions on older people and their mobilities. Transitions may occur as a result of aging (for example hearing loss or stopping child care) and have an affect on mobilities (Gilroy et al.). The key life events that Klinger discusses may be the result of decisions taken in anticipation of limitations to mobilities in older age. On the other hand the limitations to mobilities that returning immigrants experience, according to Kalaw in Chapter 12, seem to come as something of a surprise and certainly a disappointment: Filipino males who have spent many years working abroad and providing for family back home in the Philippines, return home to find that they are less able to pick up previous (generally physical) occupations due to diminished physical fitness as these occupations have not been practised for a long time.

Fourth is the co-constitution of generational mobilities and urban space. Over a 24-hour period of observing a 'shared space' in Brighton, Murray and Robertson (Chapter 8) found a shifting of generationally identified dominance: the maximum diversity of generations was observed in the middle of the day but at night time groups of young adults were dominant. The effect of the changes in generational diversity had a marked impact on the spatial practices of the shared street space, and this effect appeared to be self-perpetuating: the narrower range of generations dominating the space had a negative impact on encouraging a broader range of generations to share the space. Meanwhile Johnson in Chapter 4 describes how the relationships between mothers (and other relatives) and daughters apply pressures that move street-connected girls in Nairobi in and out of education and working on the streets. Intergenerational cultures of mobilities, in this collection, are defined in various ways: here most clearly identified in the expectations, anticipations and perceptions of those interviewed as participants of the research. These are cultures that are experienced rather than assumed or observed by others; they are immersive, and we have drawn attention to the mobilities within and around the edges of these cultures, as they change within and across generations. Effects of speed on mobilities as an aspect of culture are seen in Murray and Robertson, and elucidated in Harley's detailed examination of the modes of travel that different generations engage in while using a mobile phone. Both of these studies were made in the same 'shared space', clearly demonstrating how the slowed down nature of the space allows for a different kind of scrutiny.

A pervading theme throughout is the embodiment of age – how the body feels and moves and feels as it moves – becomes a far more complex concept when taken to stages of the lifecourse and generation. The evidence and perceptions of injustices that are captured in the chapters here are material, embodied; they are felt and affect the day-to-day lives of those identified as 'different'. Reaction and rejection and an understanding of tolerance and friction on the basis of perceived differences of mobilities expand through the focus of intergenerationality, with a number of adjacent themes emerging from the chapters, including the intersections of gender, the specific nature of relationships and networks whether inter- or extra-familial, the significance of economic, educational and health-related profiles, and the use of technologies.

Journeys through time and space are both real and imagined, and the sharing of these experiences is important in identifying where and when injustices occur so that we can take pre-emptive actions. The stories recounted in this collection draw attention to the spatial significance of mobilities, in terms of scope, scale, frequency and distance. Just as 'age' has become an increasingly ambiguous, mutable term, we contend that perceptions of the relationship between age, space and place have become very much more potent as characteristics of the lifecourse.

Bibliography

Abrahamsson, S. and Simpson, P. 2011. The limits of the body: boundaries, capacities, thresholds. *Social and Cultural Geography*, 12(4), 331–338.

Adams, E. and Ingham, S. 1998. *Changing Places: Children's Participation in Environmental Planning*. London: The Children's Society.

Adams, J. 1999. *The Social Implications of Hypermobility*. Paris: OECD.

Adey, P. 2006. If mobility is everything then it is nothing: towards a relational politics of (im)mobilities. *Mobilities*, 1(1), 75–94.

Adey, P. 2007. 'May I have your attention': airport geographies of spectatorship, position, and (im)mobility. *Environment and Planning D: Society and Space*, 25, 515–536.

Adey, P. 2009. *Mobility*. London: Routledge.

Aditjandra, P.T., Cao, X. and Mulley, C. 2012. Understanding neighbourhood design impact on travel behaviour: an application of structural equations model to a British metropolitan data. *Transportation Research Part A*, 46(1), 22–32.

Agamben, G. 1998. *Homo Sacer: Sovereign Power and Bare Life*. Stanford, CA: Stanford University Press.

Age UK. 2011. *Strengthening the Convoy: The Campaign to End Loneliness*. http:// campaigntoendloneliness.org/wp-content/uploads/downloads/2011/07/safeguarding-the-convey_-_a-call-to-action-from-the-campaign-to-end-loneliness.pdf. Accessed 21 September 2015.

Alanen, L. 2001. Explorations in generational analysis. In Alanen, L. and Mayall, B., *Conceptualizing Child-Adult Relations*. London: Routledge.

Albom, M. 1997. *Tuesdays with Morrie: An Old Man, A Young Man, and Life's Greatest Lesson*. New York: Broadway Books.

Aldred, R. and Jungnickel, K. 2014. Why culture matters for transport policy: the case of cycling in the UK. *Journal of Transport Geography*, 34, 78–87.

Alves, T. 2007. Art, light and landscape: new agendas for urban development. *European Planning Studies*, 15(9) 1247–1260. Amin, A. 2002Ethnicity and the multicultural city: living with diversity. *Environment and planning A*, 34, 959–980.

Amin, A. 2006. The good city. *Urban Studies*, 43, 1009–1023.

Amin, A. and Thrift, N. 2002. Cities – reimagining the urban Cambridge: Polity Commission for Architecture and the Built environment. In CABE (2007) *This Way to Better Streets: 10 Case Studies on Improving Street Design*. London: CABE.

Andrews, G.J., Evans, J. and Wiles, J.L. 2012. Re-spacing and re-placing gerontology: relationality and affect. *Ageing and Society*, 33(8), 1339–1373.

Andrews, G.J., Hall, E., Evans, B. and Colls, R. 2012. Moving beyond walkability: on the potential of health geography. *Social Science and Medicine*, 75(11), 1925–1932.

Antonucci, T.C., Ajrouch, K.J. and Birditt, K. 2006. Social relations in the third age. *Annual Review of Gerontology and Geriatrics*, 26, 193–210.

Anvari, B. 2012. A mathematical model for driver and pedestrian interaction in shared space environments. 44th Universities Transport Study Group Conference. Aberdeen.

Aristotle. 350 BCE-a (1994–2009). *De Anima*. Trans. J.A. Smith. Boston: MIT Classics.

Aristotle. 350 BCE-b (1994–2009). *Nicomachean Ethics*. Trans. W.D. Ross. Boston: MIT Classics.

Arthur, I.I. 2013. *Streetism: A Socio-Cultural and Pastoral Theological Study of a Youth Problem in Ghana*. Bloomington, IN: Author House. Asis, M.B. 2006. International migration and development in the Philippines. In Palme, J. and Tamas, K. (eds) *How Migration Can Benefit Development*. Stockholm: Institute for Futures Studies, 149–165.

Asis, M.B. and Piper, N. 2008. Researching international labour migration in Asia. *Sociological Quarterly*, 49, 423–444.

AWOC (Ageing Without Children). 2015. About AWOC. https://awoc.org/

Axhausen, K.W. 2008. Social networks, mobility biographies, and travel: survey challenges. *Environment and Planning B*, 35(6), 981–996.

Axisa, J.J., Newbold, K.B. and Scott, D.M. 2012. Migration, urban growth and commuting distance in Toronto's commuter shed. *Area*, 44(3), 344–355.

Ayis, S., Gooberman-Hill, R., Bowling, A. and Ebrahim, S. 2006. Predicting catastrophic decline in mobility among older people. *Age and Ageing*, 35, 382–387.

Bailey, A.J. 2009. Population geography: lifecourse matters. *Progress in Human Geography*, 33(3), 407–418.

Baker, L. and Gringart, E. 2009. Body image and self-esteem in older adulthood. *Ageing and Society*, 29(6), 977–995.

Ballagas, R., Borchers, J., Rohs, M. and Sheridan, J. 2006. The Smart Phone: a ubiquitous input device. *IEEE Pervasive Computing*, 5, 70–77.

Ballatore, M. 2007. L'expérience de mobilité des étudiants ERASMUS: les usages inégalitaires d'un programme d' 'échange'. Une comparaison Angleterre/France/Italie. Ph.D. thesis in sociology, Université Aix-Marseille and Università degli studi di Torino. http://tel.archives-ouvertes.fr/tel-00204795

Barker, J., Kraftl, P., Horton, J. and Tucker, F. 2009. The road less travelled: new directions in children's and young people's mobility. *Mobilities*, 4(1), 1–10.

Barnett, C., 2005. Ways of relating: hospitality and the acknowledgement of otherness. *Progress in Human Geography*, 29, 5–21.

Beazley, H. and Ennew, J. 2006. Participatory methods and approaches: tackling the two tyrannies. In Desai, V. and Potter, R.B., *Doing Development Research*. London: Sage.

Beckmann, J. 2005. Mobility and safety. In Featherstone, M., Thrift, N. and Urry, J. (eds) *Automobilities*. London: Sage.

Beige, S. and Axhausen, K.W. 2012. Interdependencies between turning points in life and long-term mobility decisions. *Transportation*, 39(4), 857–872.

Beltrame, L. 2007. Realtà e retorica del Brain Drain in Italia. Stime statistiche, definizioni pubbliche e interventi politici. *Quaderni del dipartimento di Sociologia e Ricerca sociale*, 35, Università di Trento.

Bentley, F., Basapur, S. and Kumar, S. 2011. Promoting intergenerational communication through location-based asynchronous video communication. UbiComp, 13th International Conference on Ubiquitous Computing. Beijing, China. 17–21 September. 3–12.

Bergmann, S. and Sager, T. 2008. In between standstill and hypermobility: introductory remarks to a broader discourse. In Bergmann, S. and Sager, T. (eds) *The Ethics of Mobilities: Rethinking Place, Exclusion, Freedom and Environment*. Aldershot: Ashgate.

Berk, L. 2004. *Development Through the Lifespan*. Boston: Pearson Education.

Bhugra, D. and Becker, M.A. 2005. Migration, cultural bereavement and cultural identity. *World Psychiatry*, 4(1), 18–24.

Biggs, S. 1997. Choosing not to be old? Masks, bodies and identity management in later life. *Ageing and Society*, 17(5), 553–570.

Binnie, J., Holloway, J., Millington, S. and Young, C. (eds) 2006. *Cosmopolitan Urbanism*. London: Routledge.

Bissell, D. 2013. Pointless mobilities: rethinking proximity through the loops of neighbourhood. *Mobilities*, 8(3), 349–367.

Bissell, D., Adey, P. and Laurier, E. 2011. Introduction to the special issue on Geographies of the Passenger. *Journal of Transport Geography*, 19(5), 1007–1009.

Bissell, D. and Fuller, G. 2011. Stillness unbound. In Bissell, D. and Fuller, G. (eds) *Stillness in a Mobile World*. London: Routledge.

Black, S.E. and Devereux, P.J. 2010. Recent developments in intergenerational mobility. *Working paper 15889*. Cambridge, MA: National Bureau of Economic Research.

Blit-cohen, E. and Litwin, H. 2004. Elder participation in cyberspace: a qualitative analysis of Israeli retirees. *Journal of Aging Studies*, 18, 385–398.

Blomberg, J., Burrell, M. and Guest, G. 2003. An ethnographic approach to design. In Jacko, J.A. and Sears, A. (eds) *The Human Computer Interaction Handbook*. Mahwah, NJ: Lawrence Erlbaum Associates.

Borden, I. 2001. Another pavement, another beach: skateboarding and the performative critique of architecture. In Borden, I., Kerr, J. and Rendell, J. with Pivaro, A. (eds) *The Unknown City: Contesting Architecture and Social Space*. Cambridge MA and London: MIT Press.

Bostock, L. 2001. Pathways of disadvantage? Walking as a mode of transport among low-income mothers. *Health and Social Care in the Community*, 9, 11–18.

Bourn, R. 2013. *No Entry! Transport Barriers Facing Young People*. Available at: www.bettertransport.org.uk/files/No_Entry_final_report_definitive_0.pdf

Bovenkerk, F. and Research Group for European Migration Problems. 1974. *The Sociology of Return Migration: A Bibliographic Essay*. The Hague: Martinus Nijhoff.

Bratzel, S. 1999. Conditions of success in sustainable urban transport policy: policy change in 'relatively successful' European cities. *Transport Reviews*, 19(2), 177–190.

Braun, V. and Clarke, V. 2006. Using thematic analysis in psychology. *Qualitative Research in Psychology*, 3, 77–101.

Brettell, C.B. 1988. Emigration and household structure in a Portuguese parish, 1850–1920. *Journal of Family History*, 13(2), 33–57.

Bristol City Council. 2013. *Neighbourhood Partnership Statistical Profile 2013: Avonmouth Kingsweston*. Bristol: Bristol City Council.

British Youth Council and Youth Select Committee. 2012. *Transport and Young People*. London: British Youth Council and Youth Select Committee.

Brush, A.J.B., Inkpen, K.M. and Tee, K. 2008. SPARCS: exploring sharing suggestions to enhance family connectedness. *CSCW Journal*, 629–638.

Buehler, R. 2011. Determinants of transport mode choice: a comparison of Germany and the USA. *Journal of Transport Geography*, 19(4), 644–657.

Burawoy, M. 1976. The functions and reproduction of migrant labour: comparative material from Southern Africa and the United States. *American Journal of Sociology*, 5, 1050–1086.

Busch-Geertsema, A. and Lanzendorf, M. 2015. Mode decisions and context change: what about the attitudes? A conceptual framework. In Attard, M. and Shiftan, Y. (eds) *Sustainable Urban Transport*. Bradford: Emerald Group, 23–42.

Bynner, J. and Wadsworth, M. 2011. Generation and change in perspective. In *A Companion to Life Course Studies: The Social and Historical Context of the British Birth Cohort Studies*. London and New York: Routledge, 203–221.

Calasanti, T. 2005. Ageism, gravity, and gender: experiences of aging bodies. *Generations*, 29(3), 8–12.

Campbell, S. 2007. Perceptions of mobile phone use in public settings: a cross-cultural comparison. *International Journal of Communication*, 1, 738–757.

Cantle, E. 2004. *The End of Parallel Lives? The Report of the Community Cohesion Panel*. London: Home Office.

Cao, X., Mokhtarian, P.L. and Handy, S.L. 2007. Do changes in neighborhood characteristics lead to changes in travel behavior? A structural equations modeling approach. *Transportation*, 34(5), 535–556.

Cao, X., Mokhtarian, P.L. and Handy, S.L. 2009. Examining the impacts of residential self-selection on travel behaviour: a focus on empirical findings. *Transport Reviews*, 29(3), 359–395.

Caragliu, A., Del Bo, C. and Nijkamp, P. 2011. Smart cities in Europe. *Journal of Urban Technology*, 18(2), 65–82.

Cassarino, J.P. 2004. Theorising return migration: the conceptual approach to return migrants revisited. *International Journal of Multicultural Societies*, 6(2), 253–279.

Castel, R. 2003. *From Manual Workers to Wage Laborers: Transformation of the Social Question*. Piscataway, NJ: Transaction.

Castles, S. 2010. Understanding global migration: a social transformation perspective. *Journal of Ethnic and Migration Studies*, 36(10), 1565–1586.

Cerase, F.P. 1967. Study of Italian migrants returning from the USA. *International Migration Review*, 1, 67–74.

Cervero, R. 1998. *The Transit Metropolis: A Global Inquiry*. Washington, DC: Island Press.

Cervero, R. and Kockelman, K. 1997. Travel demand and the 3Ds: density, diversity, and design. *Transportation Research Part D*, 2(3), 199–219.

Chapin, S. 1974. *Human Activity Patterns in the City: Things People Do in Time and Space*. New York: Wiley.

Charmaz, K. 2006. *Constructing Grounded Theory: A Practice Guide Through Qualitative Analysis*. London: Sage.

Chatman, D.G. 2009. Residential choice, the built environment, and nonwork travel: evidence using new data and methods. *Environment and Planning A*, 41(5), 1072–1089.

Chatterjee, K., Sherwin, H. and Jain, J. 2013. Triggers for changes in cycling: the role of life events and modifications to the external environment. *Journal of Transport Geography*, 30, 83–193.

Chua, P.-H., Jung, Y., Lwin, M.O. and Theng, Y.-L. 2013. Let's play together: effects of video-game play on intergenerational perceptions among youth and elderly participants. *Computers in Human Behavior*, 29, 2303–2311.

Clark, B., Chatterjee, K. and Lyons, G. 2015. Towards a theory of the dynamics of household car ownership: insights from a mobility biographies approach. In Scheiner, J. and Holz-Rau, C. (eds) *Mobility Biographies and Mobility Socialisation*. London: Springer.

Clark, B., Chatterjee, K., Melia, S., Knies, G. and Laurie, H. 2014. Life events and travel behaviour: exploring the interrelationship using UK household longitudinal study data. *Transportation Research Record*, 2413, 54–64.

Clark, W.A.V. and Withers, S.V. 2007. Family migration and mobility sequences in the united states: spatial mobility in the context of the life course. *Demographic Research*. Web. Accessed 11 April 2015. www.demographic-research.org/volumes/vol17/20/17-20.pdf

Classen, C. 2006. Aeromobility and work. *Environment and Planning A*, 38, 301–312.

Cloke, P. 2002. Deliver us from evil? Prospects for living ethically and acting politically in human geography. *Progress in Human Geography*, 26(5), 587–604.

Conci, M., Pianesi, F. and Zancanaro, M. 2010. Useful, social and enjoyable: mobile phone adoption by older people. In Gross, T., Gulliksen, J. and Kotze, P. (eds) *INTERACT 2009: Proceedings of the 12th IFIP TC 13 International*. Berlin: Springer, 1–14.

Conradson, D. and Latham, A. 2005. Transnational urbanism: attending to everyday practices and mobilities. *Journal of Ethnic and Migration Studies*, 31(2), 227–233.

Cornejo, R., Tentori, M. and Favela, J. 2013. Enriching in-person encounters through social media: a study on family connectedness for the elderly. *International Journal of Human-Computer Studies*, 71(9), 889–899.

Cornwall, A. 2004. Spaces for transformation? Reflections of power and difference in participation in development. In Hickey, S. and Mohan, G. (eds) *Participation: From Tyranny to Transformation? Exploring New Approaches to Participation in Development*. London: Zed Books.

Corti, P. 2003. L'emigrazione italiana in Francia: un fenomeno di lunga durata. *Altre Italie*, 26, 4–25.

Cosgrove, D. 2011. Prologue: geography within the humanities. In Daniels, S., DeLyser, D., Entrikin, J.N. and Richardson, D. (eds) *Envisioning Landscapes, Making Worlds: Geography and the Humanities*. London: Routledge.

Cowley, C. 2013. The last chapter in the story: a place for Aristotle's *Eudaemonia* in the lives of the terminally ill. *Online Journal of Health Ethics*, 3(1), 1–10.

Crenshaw, K. 1989. Demarginalizing the intersection of race and sex: a black feminist critique of antidiscrimination doctrine, feminist theory and antiracist politics. *University of Chicago Legal Forum*, 140, 139–167.

Cresswell, T. 2006. *On the Move: Mobility in the Modern Western World*. London: Routledge.

Cresswell, T. 2010. Towards a politics of mobility. *Environment and Planning D: Society and Space*, 28, 17–31.

Cresswell, T., 2011a. Race, mobility and the humanities: a geosophical approach. In Daniels, S., DeLyser, D., Entrikin, J.N. and Richardson, D. (eds) *Envisioning Landscapes, Making Worlds: Geography and the Humanities*. New York and London: Routledge.

Cresswell, T. 2011b. Mobilities I: catching up. *Progress in Human Geography*, 35(4), 550–558.

Cucchiarato, C. 2011. Guerra di cifre: perchè è cosi difficile capire chi e quanti sono gli Italiani all'estero? *Altreitalie*, 43, 64–72.

Daniels, S., DeLyser, D., Entrikin, J.N. and Richardson, D. (eds) 2011. *Envisioning Landscapes, Making Worlds: Geography and the Humanities*. London: Routledge.

Dannefer, D. 2003. Toward a global geography of the life course: challenges of late modernity to the life course perspective. In Mortimer, J.T. and Shanahan, M. (eds) *Handbook of the Life Course*. New York: Kluwer.

Dannefer, D. and Uhlenberg, P. 1999. Paths of the life course: a typology. In V.L. Bengtson and K.W. Schaie (eds) *Handbook of Theories of Aging: In Honor of Jim Birren*. New York: Springer.

Davis, M. 2003. Intergenerational practice: an idea whose time has come. *Gerontologist*, 43, 287.

De Certeau, M. 1984. Walking in the city. In *The Practice of Everyday Life*. Berkeley: University of California Press.

de Medeiros, K. B. 2014. *Narrative Gerontology in Research and Practice*. New York: Springer.

Dean, M. 1999. *Governmentality: Power and Rule in Modern Society*. Thousand Oaks CA, London and New Delhi: Sage.

Dean, M. 2002. Powers of life and death beyond governmentality. *Cultural Values*, 6 (1–2), 119–138.

Dear, M. 2011. Creativity and place. In Dear, M., Ketchum, J., Luria, S. and Richardson, D. (eds) *GeoHumanities: Art, History, Text at the Edge of Place*. New York and London: Routledge, 9–18.

Deffner, J., Götz, K., Schubert, S., Potting, C., Stete, G., Tschann, A. and Loose, W. 2006. *Schlussbericht zu dem Projekt 'Nachhaltige Mobilitätskultur'. Entwicklung eines integrierten Konzepts der Planung, Kommunikation und Implementierung einer nachhaltigen, multioptionalen Mobilitätskultur. Projekt 70.0749/04(FOPS) BMVBS Referat A 32*. Frankfurt am Main.

Degen, M. and Rose, G. 2012. The sensory experiencing of urban design: the role of walking and perceptual memory. *Urban studies*, 49(15), 3271–3287.

Del Boca, D. and Rosina, A. 2009. *Famiglie Sole. Sopravivere con un Welfare inefficiente*. Bologna: Il Mulino.

Delbosc, A. and Currie, G. 2013. Causes of youth licensing decline: a synthesis of evidence. *Transport Reviews*, 33(3), 271–290.

Del Prà, A. 2011. Giovani Italiani a Berlino: nuove forme di mobilità europea. *Altreitalie*, 43, 103–125.

Department for Transport. 2007. *Manual for Streets*. London: Thomas Telford Publishing.

Department for Transport. 2009. *DfT Shared Space Project Stage 1: Appraisal of Shared Space*. London: DfT.

Department for Transport. 2011. *National Travel Survey*. https://www.gov.uk/governm ent/statistics/national-travel-survey-2010 (Accessed 18 September 2015).

dePoy, E. and Gitlin, L. 2015. *Introduction to Research: Multiple Strategies for Health and Human Services*. 3rd edn. St Louis, MO: Mosby.

Devereux, S. and Sabates-Wheeler, R. 2004. Transformative social protection. IDS Working Paper 232. University of Sussex, Brighton: Institute of Development Studies.

Devereux, S. and Sabates-Wheeler, R. 2011. Transformative social protection for Africa's children. In Handa, S., Devereux, S. and Webb, D. (eds) *Social Protection for Africa's Children*. London: Routledge.

DeWalt, K. and DeWalt, B. 2012. *Participant Observation. A Guide for Fieldworkers.* Plymouth: Altamira Press.

Dickinson, A. and Hill, R.L. 2007. Keeping in touch: talking to older people about computers and communication. *Educational Gerontology*, 33(8), 613–630.

Diminescu, D., Berthomière, W. and Ma Mung, E. 2014. Les traces de la dispersion. *Revue Européenne des Migrations Internationales*, 30, 3–4.

Döring, L., Albrecht, J., Scheiner, J. and Holz-Rau, C. 2014. Mobility biographies in three generations: socialization effects on commute mode choice. *Transportation Research Procedia*, 1(1), 165–176.

Doughty, K. and Murray, L. 2014. Discourses of mobility: institutions, everyday lives and embodiment. *Mobilities*, 9, 1–20.

Dowling, R. 2000. Cultures of mothering and car use in suburban Sydney: a preliminary investigation. *Geoforum*, 31(3), 345–353.

Duncan, M. 2015. *'You Make Your Own Shed, It Is Not Something You Are Given': Learning about Community Capacity Building from the Spread of Men's Sheds in Scotland.* Report for the Joint Improvement Team by md consulting. Available at: www.jits cotland.org.uk/wp-content/uploads/2015/02/Community-Capacity-Building-Mens-Sheds-in- Scotland-Report-February-2015.pdf (Accessed 21 September 2015).

Durick, J., Robertson, T., Brereton, M., Vetere, F. and Nansen, B. 2013. Dispelling ageing myths in technology design. 25th Australian Computer-Human Interaction Conference, OzCHI 2013. Adelaide, SA, Australia. 467–476.

Dykstra, P.A. and Van Wissen, L.J.G. 1999. Introduction: the life course approach as an interdisciplinary framework for population studies. *Population Issues: An Interdisciplinary Focus.* New York: Plenum.

Edensor, T. 2010. Walking in rhythms: place, regulation, style and the flow of experience. *Visual Studies*, 25(1), 69–79.

Edmonston, B. 2013. Lifecourse perspectives on immigration. *Canadian Studies in Population*, 40, (1–2), 1–8.

Elder, G. 1975. Age differentiation and the life course. *Annual Review of Sociology*, 1, 165–190.

Elder, Jr, G. 1985a. Preface. *Life Course Dynamics: Trajectories and Transitions, 1968–1980.* Ithaca, NY: Cornell University Press.

Elder, Jr, G. 1985b. Perspective on the life course. *Life Course Dynamics: Trajectories and Transitions, 1968–1980.* Ithaca, NY: Cornell University Press.

Elder, Jr, G. 1994. Time, human agency, and social change: perspective on the life course. *Social Psychology Quarterly*, 57(1), 4–15.

Emanuel, L. 2013. Fair fares: new First bus-ticket prices welcomed. *Bristol Post.* Available at: www.bristolpost.co.uk/Fair-fares-New-bus-ticket-prices-welcomed/story-20033120-detail/story.html

Engwicht, D. 2005. *Mental Speed Bumps: The Smarter Way to Tame Traffic.* Annadale, NSW: Envirobooks.

Ennew, J. 1994. *Street and Working Children: A Guide to Planning.* Development Manual 4. London: Save the Children UK.

Enriquez, V.G. 1975. Mga batayan ng sikolohiyang Pilipino sa kultura at kasaysayan [The bases of Filipino psychology in culture and history]. *General Education Journal*, 29, 61–88.

Enriquez, V.G. 1985. *Kapwa*: a core concept in Filipino social psychology. In Aganon, D. (ed.), *Sikolohiyang Pilipino: Isyu, Pananaw at Kaalaman.* Quezon City: National Bookstore.

Erikson, E.H. 1963. *Childhood and Society*. New York: Norton.

Evans, B. 2008. Geographies of youth/young people. *Geography Compass*, 2, 1659–1680.

Feltham, F., Vetere, F. and Wensveen, S. 2007. Designing tangible artefacts for playful interactions and dialogues. Conference paper, Designing Pleasurable Products and Interfaces. Helsinki, Finland.

Fernández-Ardèvol, M. and Ivan, L. 2013. Older people and mobile communication in two European contexts. *Romanian Journal of Communication and Public Relations*, 15(3), 83–101.

Ferreira, S.M., Sayago, S. and Blat, J. 2014. Towards iTV services for older people: exploring their interactions with online video portals in different cultural backgrounds. *Technology and Disability*, 26, 199–209.

Fetterman, D. 2010. *Ethnography: Step-by-Step*. Applied Social Research Methods series, volume 17. London: Sage.

Fincham, B., McGuiness, M. and Murray, L. (eds) 2010. *Mobile Methodologies*. London: Palgrave Macmillan.

Fisker, C. 2011a. End of the road? Loss of (auto)mobility among seniors and their altered mobilities and networks: a case study of a car-centred Canadian city and a Danish city. Ph.D. dissertation, Aalborg University.

Fisker, C. 2011b. Glimpses of motility of the networked self across the life course. In Vannini, P., Budd, L., Jensen, O.B., Fisker, C. and Jiron, P. (eds) *Technologies of Mobility in the Americas*. New York: Peter Lang.

Flourish Over 50. 2014. Midlife reinventions to experience life to the fullest! Feature: How not to look old. www.flourishover50.com/tag/how-not-to-look-old/ (Accessed 2 March 2014).

Foucault, M. 1976. *The History of Sexuality, Volume 1: An Introduction*. London: Penguin.

Foucault, M. 1980. Questions on geography: an interview with the editors of the journal *Herodote*. In *Power/Knowledge: Selected Interviews and Other Writings 1972–1977*, ed. C. Gordon. Brighton: Harvester Press, 63–77.

Foucault, M. 1982. The subject and power. *Critical Inquiry*, 8(4), 777–795. Pages 777–785 written in English by Foucault; balance of 785–795 translated from the French by Lesley Sawyer.

Foucault, M. 1986. Of other spaces. *Diacritics*, 16(1), 22–27.

Foucault, M. 1991. Governmentality. In Burchell, G., Gordon, C. and Miller, P. (eds) *The Foucault Effect: Studies in Governmentality*. London: Harvester Wheatsheaf, 87–104.

Frändberg, L. 2008. Paths in transnational time-space: representing mobility biographies of young Swedes. *Geografiska Annaler, Series B*, 90(1), 17–28.

Frank, K.I. 2006. The potential of youth participation in planning. *Journal of Planning Literature*, 20(4), 351–371.

Freudendal-Pedersen, M. 2009. *Mobility in Daily Life: Between Freedom and Unfreedom*. Burlington, VT: Ashgate.

Fristedt, S., Dahl, A., Wretstrand, A., Bjorklund, A. and Falkmer, T. 2014. Changes in community mobility in older men and women: a 13-year prospective study. *PLoS ONE*, 9(2), e87827. doi:10.1371/journal.pone.0087827.

Fuchsberger, V., Neureiter, K., Sellner, W. and Tscheligi, M. 2011. Attributes of successful intergenerational online activities. In *Proceedings of the 8th International Conference on Advances in Computer Entertainment Technology – ACE '11*. Lisbon, Portugal: ACM Press, 1–8.

Gagliardi, C., Spazzafumo, L., Marcellini, F., Mollenkopf, H., Ruopilla, I., Tacken, M. and Szemann, Z. 2007. The outdoor mobility and leisure activities of older people in five European countries. *Ageing and Society*, 27, 683–700.

Gans, D. and Kiesler, S. 2001. Blurring the boundaries: cell phones, mobility, and the line between work and personal life. In Brown, B., Green, N. and Harper, R. (eds) *Wireless World: Social and Interactional Aspects of the Mobile Age*. London: Springer.

Gans, D., Putneay, N., Bengston, V. and Silverstein, M. 2009. The future of theories of ageing. In Bengston, V., Silverstein, M., Putneay, N. and Gans, D. (eds) *Handbook of Theories of Ageing*. New York: Springer.

Gardner, K. 2002. *Age, narrative and migration: the life course and life histories of Bengali elders in London*. Oxford and New York: Berg.

Gawande, A. 2014. *Being Mortal: Illness, Medicine and What Matters in the End*. London: Wellcome Foundation.

Geertz, C. 1973. Thick description: toward an interpretive theory of culture. In *The Interpretation of Cultures: Selected Essays*. New York: Basic Books, 3–30.

Gehl Architects. 2010. Paving the way for city change. http://gehlarchitects.com/cases/new-road-brighton-uk/

Geist, C. and McManus, P.A. 2008. Geographic mobility over the life course: motivations and implications. *Population, Space, and Place*, 14(4), 283–303.

Gergen, K.J. 2002. The challenge of absent presence. In Katz, J.E. and Aakhus, M. (eds) *Perpetual Contact*. Cambridge: Cambridge University Press.

Gibson, L., Moncur, W., Forbes, P., Arnott, J. and Martin, C. 2010. Designing social networking sites for older adults. In BSC HCI 2010, Dundee, 186–194.

Giele, J.Z. and Elder, Jr, G. (eds) 1998. *Methods of Life Course Research: Qualitative and Quantitative Approaches*. Thousand Oaks, CA: Sage.

Giles, H. and Gasiorek, J. 2011. Intergenerational communication practices. In Schaie, K.W. and Willis, S.L., *Handbook of the Psychology of Aging*. Amsterdam: Elsevier.

Gilleard, C. and Higgs, P. 2000. *Cultures of Ageing: Self, Citizen and the Body*. Harlow: Pearson.

Gjergji, I. (ed.) 2015. *La nuova emigrazione italiana. Cause, mete e figure sociali*. Venice: Edizioni Ca'Foscari.

GLA (Greater London Authority). 2008. *Way to Go! Planning for Better Transport*. London: Greater London Authority.

Glaser, W.A. and Habers, C. 1974. The migration and return of professionals. *International Migration Review*, 8(Summer), 227–244.

Gmelch, G. 1980. Return migration. *Annual Review of Anthropology*, 9, 135–159.

Goetzke, F. 2008. Effects in public transit use: Evidence from a spatially autoregressive mode choice model for New York. *Urban Studies*, 45(2), 407–417.

Goetzke, F. and Rave, T. 2011. Bicycle use in Germany: explaining differences between municipalities with social network effects. *Urban Studies*, 48(2), 427–437.

Goffman, E. 1961. *Asylums*. New York: Anchor Books.

Goffman, E. 1966. *Behaviour in Public Spaces*. New York: Doubleday.

Gollwitzer, P.M. 1996. The volitional benefits of planning. *The Psychology of Action: Linking Cognition and Motivation to Behavior*. New York: Guilford Press.

Golombek, S.B. 2006. Children as citizens. *Journal of Community Practice*, 14(1–2), 11–30.

Goodman, A., Jones, A., Roberts, H., Steinbach, R. and Green, J. 2012. 'We can all just get on a bus and go': rethinking independent mobility in the context of the universal provision of free bus travel to young Londoners. *Journal of Epidemiology and Community Health*, 66, 39–40.

Götz, K. and Deffner, J. 2009. Eine neue Mobilitätskultur in der Stadt – praktische Schritte zur Veränderung. In BMVBS – Bundesministerium für Verkehr, Bau und Stadtentwicklung (ed.) *Urbane Mobilität: Verkehrsforschung des Bundes für die kommunale Praxis*. Bremerhaven: NW-Verlag.

GRAB (Groupe de reflexion sur l'approche biographique). 1999. *Biographies d'enquêtes, Bilan de 14 collectes biographiques*. Paris: INED, Méthodes et Savoirs.

Graham, S. 2002. Flow city: networked mobilities and the contemporary metropolis. *Journal of Urban Technology*, 9(1), 1–20.

Granié, M.-A. and Papafava, E. 2011. Gender stereotypes associated with vehicle driving among French preadolescents and adolescents. *Transportation Research Part F: Traffic Psychology and Behaviour*, 14(5), 341–353.

Green, L. 2010. *Understanding the Life Course: Sociological and Psychological Perspectives*. Cambridge: Polity.

Gregor, P. and Newell, A.F. 2001. Designing for dynamic diversity: making accessible interfaces for older people. In WUAUC. Alcácer do Sal, Portugal, 90–92.

Grenier, A. 2012. *Transitions and the Lifecourse: Challenging the Constructions of 'Growing Old'*. Bristol: Policy Press.

Gropas, R. and Triandafyllidou, A. 2013. *Emigrating in Times of Crisis: Survey Report, Global Governance Programme*. http://globalgovernanceprogramme.eui.eu/wp-content/uploads/2014/03/SURVEY-REPORT-Emigrating-in-times-of-crisis.pdf

Guiver, J.W. 2007. Modal talk: discourse analysis of how people talk about bus and car travel. *Transportation Research Part A: Policy and Practice*, 41(3), 233–248.

Habuchi, I. 2005. Accelerating reflexivity. In Ito, M., Okabe, D. and Matsuda, M. (eds) *Personal, Portable, Pedestrian: Mobile Phones in Japanese Life*. Cambridge, MA: MIT Press.

Hacfeli, U. 2005. Public transport in Bielefeld (Germany) and Berne (Switzerland) since 1950: a comparative analysis of efficiency, effectiveness and political background. *European Journal of Transport and Infrastructure Research*, 5(3), 219–238.

Hagestad, G.O. 1998. Towards a society for all ages: new thinking, new language, new conversations. *Bulletin on Aging*, 2/3, 7–13.

Hagestad, G. and Uhlenberg, P. 2005. The social separation of old and young: a root of ageism. *Journal of Social Issues*, 61, 343–360.

Hall, E.T. 1966. *The Hidden Dimension*. New York: Anchor Books.

Hall, S. 1993. Culture, community, nation. *Cultural Studies*, 7(3), 349–363.

Hamilton-Baillie, B. 2008a. Shared space: reconciling people, places and traffic. *Built Environment*, 34(2), 161–181.

Hamilton-Baillie, B. 2008b. Towards shared space. *Urban Design*, 13, 130–138.

Hammad, S. 2011. Senses of place in flux: a generational approach. *International Journal of Sociology and Social Policy*, 31(9), 555–568.

Hammersley, M. and Atkinson, P. 2007. *Ethnography: Principles in Practice*. London: Routledge.

Hammond, V. and Musselwhite, C. 2012. The attitudes, perceptions and concerns of pedestrians and vulnerable road users to shared space: a case study. *UK Journal of Urban Design*, 18(1), 78–97.

Handy, S., Cao, X. and Mokhtarian, P.L. 2005. Correlation or causality between the built environment and travel behavior? Evidence from Northern California. *Transportation Research Part D*, 10(6), 427–444.

Hannam, K., Sheller, M. and Urry, J. 2006. Editorial: mobilities, immobilities and moorings. *Mobilities*, 1(1), 1–22.

Hanson, K. and Nieuwenhuys, O. (eds) 2013. *Reconceptualizing Children's Rights in International Development: Living Rights, Social Justice and Translations.* New York: Cambridge University Press.

Hareven, T.K. 2000. *Families, History and Social Change: Life-Course and Cross-Cultural Perspectives.* Boulder, CO: Westview Press.

Hassan, H. and Nasir, M.H.N. 2008. The use of mobile phones by older adults: A Malaysian study. *SIGACCESS Access Computing*, 92, 11–16.

Haustein, S. and Siren, A. 2014. Seniors' unmet mobility needs: how important is a driving license? *Journal of Transport Geography*, 41, 45–52.

Hawkins, H. 2011. Dialogues and doings: sketching the relationships between geography and art. *Geography Compass*, 5(7), 464–478.

Hawkins, L.C. and Lomask, M. 1956. *The Man in the Iron Lung.* Garden City, NJ: Doubleday.

Heath, S., Brooks, R., Cleaver, E. and Ireland, E. 2009. Qualitative interviewing. In *Researching Young People's Lives.* London: Sage.

Heckhausen, H. 1991. *Motivation and Action.* New York: Springer.

Heine, H., Mautz, R. and Rosenbaum, W. 2001. *Mobilität im Alltag: Warum wir nicht vom Auto lassen.* Frankfurt am Main: Campus.

Henseler, C. 2013. Introduction: generation X goes global – tales of accelerated cultures. In Henseler, C. (ed.) *Generation X Goes Global: Mapping a Youth Culture in Motion.* London: Routledge.

Heyl, B.S. 2001. Ethnographic interviewing. In Atkinson, P., Coffey, A., Delamont, S., Lofland, J. and Lofland, L. (eds) *Handbook of Ethnography.* London: Sage.

Hillman, M., Adams, J. and Whitelegg, J. 1990. *One False Move: A Study of Children's Independent Mobility.* London: Policy Studies Institute.

Hjorthol, R. 2013. Transport resources, mobility and unmet transport needs in old age. *Ageing and Society*, 33(7), 1190–1211.

Hjorthol, R.J., Levin, L. and Sirén, A.K. 2010. Mobility in different generations of older persons: the development of daily travel in different cohorts in Denmark, Norway and Sweden. *Journal of Transport Geography*, 18(5), 624–633.

Hockey, J. 2009. The life course anticipated: gender and chronologisation among young people. *Journal of Youth Studies*, 12(2), 227–241.

Hodgson, F. and Turner, J. 2003. Participation not consumption: the need for new participatory practices to address transport and social exclusion. *Transport Policy*, 10(4), 265–272.

Hoelscher, S. and Alterman, D. 2004. Memory and place: geographies of a critical relationship. *Social and Cultural Geography*, 5(3), 347–355.

Holland, C., Clark, A., Katz, J. and Peace, S. 2007. *Social Interactions in Urban Public Places.* York: Joseph Rowntree Foundation.

Homans, G. 1958. Social behavior as exchange. *American Journal of Sociology*, 63, 597–606.

Hopkins, D. and Stephenson, J. 2014. Generation Y mobilities through the lens of energy cultures: a preliminary exploration of mobility cultures. *Journal of Transport Geography*, 38, 88–91.

Hopkins, P. and Pain, R. 2007. Geographies of age: thinking relationally. *Area*, 39(3), 287–294.

Hopkins, P.E. 2006. Youthful Muslim masculinities: gender and generational relations. *Transactions of the Institute of British Geographers*, 31, 337–352.

Hugo, G. 2004. International migration in the Asia-Pacific region: emerging trends and issues. *International Migration: Prospects and Policies in a Global Market.* Oxford: Oxford University Press.

Hutchison, E. 2005. The life course perspective: a promising approach for bridging the micro and macro worlds for social workers. *Families in Society*, 86, 143–152.

Hyman, I.E., Boss, S.M., Wise, B.M., McKenzie, K.E. and Caggiano, J.M. 2010. Did you see the unicycling clown? Inattentional blindness while walking and talking on a cell phone. *Applied Cognitive Psychology*, 24(5), 597–607.

ISTAT. 2011. *Trasferimenti di residenza 2009.* Statistical report. Rome.

ITU. 2014. *World Telecommunication/ICT Indicators Database. International Telecommunication Union.* Available from www.itu.int/ITU-D/ict/statistics/ (Accessed 10 December 2014).

Ivanova, K. and Dykstra, P. 2015. *Aging Without Children. Public Policy and Aging Report.* Washington, DC: Gerontological Society of America.

Jacobs, J. 1961. *The Death and Life of Great American Cities.* New York: Vintage.

Jazwinska, E. and Okólski, M. (eds) 1996. Causes and consequences of migration. In *Central and Eastern Europe. Podlasie and Slask Opolski: Basic Trends in 1975–1994.* Warsaw: Instytut Studiow Spolecznych/Friedrich Ebert Stiftung.

Jensen, A. 2013. Mobility regimes and borderwork in the European Community. *Mobilities* 8(1), 35–51.

Jensen, O.B. 2009. Flows of meaning, cultures of movements:– urban mobility as meaningful everyday life practice. *Mobilities*, 4(1), 139–158.

Jensen, O.B. 2010. Erving Goffman and everyday life mobility. In Jacobsen, M.H. (ed.) *The Contemporary Goffman.* New York and London: Routledge.

Jensen, O.B. 2013. *Staging Mobilities.* London: Routledge.

Jensen, O., Sheller, M. and Wind, S. 2015. Together and apart: affective ambiences and negotiation in families' everyday life and mobility. *Mobilities*, 10(3), 363–382.

Jimenez, M.C. 1983. Masculinity or femininity concepts of the Filipino man and woman. In Antonio, L.F.Samson, L.L.Reyes, E.S. and Paguio, M.A. (eds) *Ulat ng ikalawang pambansang kumperensya sa sikolohiyang Pilipino* [Proceedings of the second national conference on Filipino psychology], Quezon City, Philippines. Pambansang Samahan sa Sikolohiyang Pilipino. 91–100.

Jirón, M.P. and Muñoz, L.I. 2014. Travelling the journey: understanding mobility. trajectories by recreating research paths. In Murray, L. and Upstone, S., *Researching and Representing Mobilities.* Basingstoke: Palgrave Macmillan.

Johansson, M. 2005. Childhood influences on adult travel mode choice. In Underwood, G. (ed.) *Traffic and Transport Psychology: Theory and Application.* Oxford: Elsevier.

Johnson, J.E. 1995. Rural elders and the decision to stop driving. *Journal of Community Health Nursing*, 12, 131–138.

Johnson, J.E. 1998. Older rural adults and the decision to stop driving: The influence of family and friends. *Journal of Community Health Nursing*, 15, 205–216.

Johnson, V. and Nurick, R. 2003. Developing coding systems to analyse difference. *PLA Notes*, 47, 19–24 (London: IIED).

Johnson, V., Hart, R. and Colwell, J. (eds) 2014, *Steps to Engaging Young Children in Research: The Guide and The Toolkit.* The Hague: Bernard van Leer Foundation.

Johnson, V., Johnson, L., Boneface, O., Walker, D. and Kiwanuka, A. 2015. *UNGEI Case Study Report: The Role of Girl's Education in Pendekezo Letu's Interventions in Nairobi.* New York: UNGEI.

Johnson, V., Leach, B., Beardon, H., Covey, M. and Miskelly, C. (2013). Love, sexual rights and young people: learning from our peer educators. In *How to Be a Youth Centred Organisation*. London: International Planned Parenthood Federation.

Johnston, L., MacDonald, R., Mason, P., Ridley, L. and Webster, C. 2000. *The Impact of Social Exclusion on Young People Moving into Adulthood*. London: Joseph Rowntree Foundation.

Jones, A., Steinbach, R., Roberts, H., Goodman, A. and Green, J. 2012. Rethinking passive transport: bus fare exemptions and young people's wellbeing. *Health and Place*, 18(3), 605–612.

Jones, A., Goodman, A., Roberts, H., Steinbach, R. and Green, J. 2013. Entitlement to concessionary public transport and wellbeing: a qualitative study of young people and older citizens in London, UK. *Social Science and Medicine*, 91, 202–209.

Jones, K. 2012. *Missing Million Policy Paper 2: Transport Barriers to Youth Unemployment*. Lancaster: The Work Foundation.

Joseph, A. and Hallman, B.C. 1998. Over the hill and far away: distance as a barrier to the provision of assistance to elderly relatives. *Soc. Sci. Med.*, 46(1), 631–639.

Judge, T.K., Neustaedter, C., Harrison, S. and Blose, A. 2011. Family portals: connecting families through a multifamily media space. In CHI 2011, Vancouver, BC, Canada, 1205–1214.

Jurilla, L. 1986. An exploratory study of the motivational system for parenthood of rural married couples. *Philippine Journal of Psychology*, 19, 5–17.

Kanayama, T. 2003. Ethnographic research on the experience of Japanese elderly people online. *New Media and Society*, 5(2), 267–288.

Kaparias, I., Bell, M.G.H., Miri, A., Chan, C. and Mount, B. 2012. Analysing the perceptions of pedestrians and drivers to shared space. *Transportation Research Part F: Traffic Psychology and Behaviour*, 15(3), 297–310.

Katz, J.E. and Aakhus, M. (eds) 2002. *Perpetual Contact*. Cambridge: Cambridge University Press.

Katz, R. and Lowenstein, A. 2010. Theoretical perspectives on intergenerational solidarity, conflict and ambivalence. In Izuhara, M. (ed.) *Ageing And Intergenerational Relations: Family Reciprocity from a Global Perspective*. Bristol: Policy Press.

Kaufmann, V. 2002. *Re-Thinking Mobility*. Aldershot: Ashgate.

Kenyon, S. 2011. Transport and social exclusion: access to higher education in the UK policy context. *Journal of Transport Geography*, 19(4), 763–771.

Kenyon, S., Lyons, G. and Rafferty, J. 2002. Transport and social exclusion: investigating the possibility of promoting inclusion through virtual mobility. *Journal of Transport Geography*, 10(3), 207–219.

Kesby, M. 2007. Spatialising participatory approaches: the contribution of geography to a mature debate. *Environment and Planning A*, 39(12), 2813–2831.

Kesselring, S. 2008. The mobile risk society. In Canzler, W., Kaufmann, V. and Kesselring, S. (eds) *Tracing Mobilities: Towards a Cosmopolitan Perspective*. Aldershot: Ashgate.

Kim, H., Monk, A., Wood, G., Blythe, M., Wallace, J. and Olivier, P. 2013. Timely present: connecting families across continents. *International Journal of Human-Computer Studies*, 71(10), 1003–1011.

King, R. 1986. Return migration and regional economic development: an overview. In King, R. (ed.) *Return Migration and Regional Economic Problems*. London: Croom Helm.

King, R. 2000. Generalizations from the history of return migration. In Ghosh, B. (ed.) *Return Migration: Journey of Hope or Despair.* Geneva: International Organization for Migration.

Kley, S. 2010. Explaining the stages of migration within a life-course framework. *European Sociological Review,* 27(4), 469–486.

Klinger, T., Kenworthy, J.R. and Lanzendorf, M. 2013. Dimensions of urban mobility cultures: a comparison of German cities. *Journal of Transport Geography,* 31, 18–29.

Klinger, T. and Lanzendorf, M. 2015. Moving between mobility cultures: what affects the travel behavior of new residents?*Transportation,* Online first.Kofman, E., Kohli, M., Kraler, A. and Schmoll, C. (eds) 2011. *Gender, Generation and the Family in International Migration.* Imiscoe: Amsterdam University Press.

Kohli, M. and Meyer, J.W. 1996. Social structure and the social construction of life stages. *Human Development,* 29, 145–149.

Koser, K. and Black, R. 1999. The end of the refugee cycle? In Black, R. and Koser, K. (eds) *The End of the Refugee Cycle? Refugee Repatriation and Reconstruction.* New York: Berghahn Books.

Kostyniuk, L.P. and Shope, J.T. 1998. *Reduction and Cessation of Driving Among Older Drivers: Focus Groups.* Ann Arbor: University of Michigan, Transportation Research Institute.

Kubik, S. 2009. Motivations for cell phone use by older Americans. *Gerontechnology,* 8(3), 150–164.

Kuhnimhof, T., Buehler, R., Wirtz, M. and Kalinowska, D. 2012. Travel trends among young adults in Germany: increasing multimodality and declining car use for men. *Journal of Transport Geography,* 24, 443–450.

Kulu, H. 2007. Fertility and spatial mobility in the life course: evidence from Austria. Max-Planck-Institut für demografische Forschung/Max Planck Institute for Demographic Research, 1 January. www.demogr.mpg.de/papers/working/wp-2005-002.pdf (Accessed 11 April 2015).

KurniawanS. 2008. Older people and mobile phones: a multi-method investigation. *International Journal of Human-Computer Studies,* 6(12), 889–901.

Lagrosa, M.E.D. 1986. Some family-related factors and personality variables affecting the adjustment of father-absent adolescents. *Philippine Journal of Psychology,* 19, 72–76.

Lahire, B. 2011. *The Plural Actor.* Cambridge: Polity Press.

Lanzendorf, M. 2002. Mobility styles and travel behavior: application of a lifestyle approach to leisure travel. *Transportation Research Record,* 1807, 163–173.

Lanzendorf, M. 2003. Mobility biographies: a new perspective for understanding travel behavior. 10th International Conference on Travel Behaviour Research, Lucerne, Switzerland.

Lanzendorf, M. 2010. Key events and their effect on mobility biographies: the case of childbirth. *International Journal of Sustainable Transportation,* 4(5), 272–292.

Lanzendorf, M. and Busch-Geertsema, A. 2014. The cycling boom in large German cities: empirical evidence for successful cycling campaigns. *Transport Policy,* 36, 26–33.

Lash, S., 2002. Foreword: individualization in a non-linear mode. In Beck, U. and Beck-Gernsheim, E., *Individualisation.* London: Sage

Laws, G. 1994. Spatiality and age relations. In Jamieson, A., Harper, S. and Victor, C., *Critical Approaches to Ageing in Later Life.* Milton Keynes: Open University Press.

Lawton, K., Cooke, G. and Pearce, N. 2014. *The Condition of Britain: Strategies for Social Renewal.* London: Institute for Public Policy Research.

Leavy, S.A. 2011. The last of life: psychological reflections on old age and death. *The Psychoanalytic Quarterly*, 80, 699–715.

Lefebvre, H. 1991. *The Production of Space*. Trans. D. Nicholson-Smith. Oxford: Blackwell.

Lefebvre, H. 2004. *Rhythmanalysis: Space, Time and Everyday Life*. Trans. S. Elden and G. Moore. London: Continuum.

Lepa, J. and Tatnall, A. 2002. The GreyPath web portal – reaching out to virtual communities of older people in regional areas. ITiRA Conference 2002, Rockhampton, Queensland, Australia, 97–103.

Life 1952. A better break for polio patients. 19 October.

Linde, C. 1993. *Life Stories: The Creation of Coherence*. Oxford: Oxford University Press.

Lindley, S.E., Harper, R. and Sellen, A. 2009. Desiring to be in touch in a changing communications landscape: attitudes of older adults. In CHI 2009, Boston, MA. ACM, 1693–1702.

Line, T., Chatterjee, K. and Lyons, G. 2012. Applying behavioural theories to studying the influence of climate change on young people's future travel intentions. *Transportation Research Part D: Transport and Environment*, 17(3), 270–276.

Ling, R., Haddon, L. and Klamer, L. 2001. The understanding and use of the internet and the mobile telephone among contemporary Europeans. ICUST 2001, Paris, France, May.

Lloyd, J. 2008. The state of intergenerational relations today: a research and discussion paper. London: International Longevity Centre.

Lord, S., Després, C. and Ramadier, T. 2011. When mobility makes sense: a qualitative and longitudinal study of the daily mobility of the elderly. *Journal of Environmental Psychology*, 31(1), 52–61.

Lucas, K. 2012. Transport and social exclusion: Where are we now? *Transport Policy*, 20, 105–113.

Lucas, K. 2004. *Transport and Social Exclusion. A Survey of the Group of Seven Nations*. London: FIA Foundation.

Luria, S. 2012. The art and science of literary geography: practical criticism in 'America's wasteland'. *American Literary History*, 24(1), 189–204.

Lynn, J. and Adamson, D.M. 2003. *Living Well at the End of Life: Adapting Health Care to Serious Chronic Illness in Old Age*. Santa Monica, CA: RAND Health.

Macrohom, J.W. 1978. Roles of husband, wife, and both husband and wife as perceived by college students. Unpublished doctoral dissertation, Centro Escolar University, Manila, Philippines.

Maddrell, A., 2013. Living with the deceased: absence, presence and absence-presence. *Cultural Geographies*, 20(4), 501–522.

Manderscheid, K., 2014. Criticising the solitary mobile subject: researching relational mobilities and reflecting on mobile methods. *Mobilities*, 9(2), 188–219.

Mann, H.S. 2012. Ancient virtues, contemporary practices: an Aristotelian approach to embodied care. *Political Theory*, 40(2), 194–221.

Mannheim, K. 1952 [1923]. The problem of generations. In Kecskemeti, P. (ed.) *Essays on the Sociology of Knowledge*. London: Routledge and Kegan Paul.

Mannion, G. 2010. After participation: the socio-spatial performance of intergenerational becoming. In Percy-Smith, B. and Thomas, N. (eds) *A Handbook of Children and Young People's Participation: Perspectives from Theory and Practice*. London: Routledge, 330–342.

Marmot, M. 2010. *Fair Society, Healthy Lives.* (The Marmot Review). London: University College London.

Marshall, S. 2009. *Cities: Design and Evolution.* London: Routledge

Marshall, M.N. 1996. The key informant techniques. *Family Practice*, 13, 92–97.

Marshall, V.W. and Mueller, M.M. 2003. Theoretical roots of the life-course perspective. *Social Dynamics of the Life Course: Sequences, Institutions, and Interrelations.* New York: Aldine de Gruyter.

Martin, F.D. and Jacobus, L.A. 1983. *The Humanities Through the Arts.* New York: McGraw-Hill.

Martin-Matthews, A. 2007. Situating 'home' at the nexus of the public and private spheres: ageing, gender and home support in Canada. *Current Sociology*, 55(2), 229–249.

Mason, M. 2003. *Breath: Life in the Rhythm of an Iron Lung – A Memoir.* Asheboro, NC: Down Home Press.

Massey, D. 2005. *For Space.* Thousand Oaks CA and London: Sage.

Massey, D., Arango, J., Hugo, G., Kouaouci, A., Pellegrino, A. and Taylor, J.E. 1993. Theories of international migration: a review and appraisal. *Population and Development Review*, 19(3), 431–466.

Matthews, H. 1992. *Making Sense of Place: Children's Understandings of Large-scale Environments.* Hemel Hempstead: Harvester Wheatsheaf.

Matthies, E., Klöckner, C.A. and Preissner, C.L. 2006. Applying a modified moral decision making model to change habitual car use: how can commitment be effective? *Applied Psychology*, 55(1), 91–106.

Mauger, G. 2010. Jeunesse: essai de construction d'objet. *Agora débats/jeunesses*, 56, 9–24.

Maxwell, J.H. 1986. The iron lung: halfway technology or necessary step? *The Milbank Quarterly*, 64(1), 3–29.

Mayer, K.U. and Tuma, N.B. 1990. *Event History Analysis in Life Course Research.* Madison: University of Wisconsin Press.

McCann-Erickson Survey Group. 1995. *Insights on the Urban Filipino Male.* Manila: Mc-Cann-Erickson Philippines.

McDonald, R. 2011. The value of art and the art of evaluation. In Bates, J. (ed.) *The Public Value of the Humanities.* London: Bloomsbury Academic.

McNeil, C. and Hunter, J. 2014. *The Generation Strain: Collective Solutions to Care in an Ageing Society.* London: IPPR.

Mead, M. 1970. *Culture and Commitment: A Study of the Generation Gap.* New York: Doubleday for the American Museum of Natural History.

Medina, B.T.G. 2001. *The Filipino Family.* 2nd edn. Quezon City, Philippines: University of the Philippines Press.

Merleau-Ponty, M. 1962. *Phenomenology of Perception.* London: Routledge and Kegan Paul.

Merriman, P. 2009. Mobility. In Kitchin, R. and Thrift, N. (eds) *International Encyclopedia of Human Geography.* Amsterdam: Elsevier, 134–143.

Metz, D. 2012. Demographic determinants of daily travel demand. *Transport Policy*, 21, 20–25,

Miciukiewicz, K. and Vigar, G. 2013. Encounters in motion: considerations of time and social justice in urban mobility research. In Henckel, D., Thomaier, S., Konecke, B., Zedda, R. and Stabilini, S. (eds) *Space-Time Design of the Public City.* Amsterdam: Springer Link.

Mikkelsen, M. and Christensen, P. 2009. Is children's independent mobility really independent? A study of children's mobility combinng ethnography and GPS/mobile phone technologies. *Mobilities*, 4(1), 37–58.

Mills, C.W. 1959. *The Sociological Imagination*. New York: Oxford University Press.

Miranda, A. 2008. Le migrazioni italiane in Francia tra trasmissione intergenerazionale, oblio e nuove mobilità. *Rapporto Italiani nel mondo 2008*. Rome: Fondation Migrantes, 316–327.

Mizen, P. and Ofosu-kusi, Y. 2013. Agency as Vulnerability: Accounting for children's movement to the streets of Accra. *Sociological Review*, 61(2), 363–382.

Mokhtarian, P.L. and Salomon, I. 2001. How derived is the demand for travel? Some conceptual and measurement considerations. *Transportation Research Part A*, 35(8), 695–719.

Mom, G. 2011. Encapsulating culture: European car travel, 1900–1940. *Journal of Tourism History*, 3(3), 289–307.

Moncrieffe, J. 2009. Introduction: intergenerational transmissions: cultivating children's agency?*IDS Bulletin*, 40(1), 1–8 (Brighton: IDS).

Monk, A., Carroll, J., Parker, S. and Blythe, M. 2004. Why are mobile phones annoying? *Behaviour and Information Technology*, 23(1), 33–41.

Mueller, S. 2013. Middling transnationalism and translocal lives: young Germans in the UK. Ph.D. thesis, University of Sussex.

Müggenburg, H., Busch-Geertsema, A. and Lanzendorf, M. 2015. Mobility biographies: a review of achievements and challenges of the mobility biographies approach and a framework for further research. *Journal of Transport Geography*, 46, 151–163.

Mulder, C.H. and Wagner, M. 1993. Migration and marriage in the life course: a method for studying synchronized events. *European Journal of Population*, 9(1). 55–76.

Muñoz, D., Cornejo, R., Ochoa, S.F., Favela, J., Gutierrez, F. and Tentori, M. 2013. Aligning intergenerational communication patterns and rhythms in the age of social media. ChileCHI. Temuco, Chile, 66–71.

Murray, L. 2009. Looking at and looking back: visualization in mobile research. *Qualitative Research*, 9(4), 469–488.

Murray, L. 2015a. Age-friendly mobilities: a transdisciplinary and intergenerational perspective. *Journal of Transport and health*, 2(2), 302–307.

Murray, L. 2015b. Rethinking children's independent mobility and revealing cultures of children's agentic and imaginative mobilities through *Emil and the Detectives*. *Transfers: Interdisciplinary Journal of Mobility Studies*, 5(1), 28–45.

Murray, L. 2017. Introduction: Conceptualising intergenerational mobilities. In Murray, L. and Robertson, S., *Intergenerational mobilities: relationality, age and the lifecourse*. London: Routledge

Murray, L. and Mand, K. 2013. Travelling near and far: placing children's mobile emotions. *Emotion, Space and Society*, 9, 72–79.

Musselwhite, C.A. and Shergold, I. 2013. Examining the process of driving cessation in later life. *European Journal of Ageing*, 10(2), 89–100.

Musselwhite, C., Walker, I. and Holland, C. 2015. Transport, travel and mobility in later life. *Journal of Transport and Health*, 2(1), 1–94.

Mwega, F.M. 2010. Kenya Phase 2. Global Financial Crisis discussion series, Paper 17. London: Overseas Development Institute.

Nasar, J.L. and Troyer, D. 2013. Pedestrian injuries due to mobile phone use in public places. *Accident Analysis and Prevention*, 57, 91–95.

Newman, O. 1996. *Creating Defensible Space*. Washington, DC: Department of Housing and Urban Development Office of Policy Development and Research.

Nicholson, S. 2011. The 0 to 100 Project. http://itunes.apple.com/us/app/0-to-100-p roject/id425167023?mt=8 (Accessed 4 October 2011).

Nicholson, C. and Hockley, J. 2011. Death and dying in older people. In Oliviere, D., Monroe, B. and Payne, S. (eds) *Death, Dying, and Social Differences*. Oxford: Oxford University Press, 101–109.

Nijs, G. and Daems, A. 2012. And what if the tangible were not, and vice versa? On boundary works in everyday mobility experience of people moving into old age. *Space and Culture*, 18(3), 186–197.

Nikander, P. 2009. Doing change and continuity: age identity and the micro-macro divide. *Ageing and Society*, 29(6), 863–881.

Nordbakke, S. 2013. Capabilities for mobility among urban older women: barriers, strategies and options. *Journal of Transport Geography*, 26, 166–174.

Nordbakke, S. and Schwanen, T. 2014. Well-being and mobility: a theoretical framework and literature review focusing on older people. *Mobilities*, 9(1), 104–129.

Nussbaum, J.F., Pecchioni, L.L., Robinson, J.D. and Thompson, T.L. 2000. *Communication and Aging*. 2nd edn. New Jersey: Lawrence Erlbaum Associates.

Oakil, A.T.M. 2013. Temporal dependence in life trajectories and mobility decisions. Ph.D. Dissertation, Utrecht University.

Oakil, A.T.M., Ettema, D., Arentze, T. and Timmermans, H. 2014. Changing household car ownership level and life cycle events: an action in anticipation or an action on occurrence. *Transportation*, 41(4), 889–904.

O'Brien, M., Jones, D., Sloan, D. and Rustin, M. 2000. Children's independent spatial mobility in the urban public realm. *Childhood*, 7(3), 257–277.

Ofcom. 2011. *Communications Market Report 2011*. London: Ofcom.

Ofcom. 2012. *Adults Media Use and Attitudes Report 2012*. London: Ofcom.

Okie, S. 2008. Home delivery: bringing primary care to the housebound elderly. *New England Journal of Medicine*, 359(23), 2409–2412.

Okolski, M. 2004. The effects of political and economic transition on international migration in Central and Eastern Europe. In Massey, D.S. and Taylor, J.E. (eds) *International Migration: Prospects and Policies in a Global Market*. Oxford: Oxford University Press.

Oleinikova, O. 2013. Life-course strategies and labor migration: Ukrainians in Italy and Poland. *Journal of National Taras Shevchenko University of Kyiv: Sociology*, 1(4), 34–41.

O'Rand, A.M. 1996. Context, selection and agency in the life course: linking social structure and biography. In Weymann, A. and Heinz, W.R. (eds) *Society and Biography: Interrelationships Between Social Structure, Institutions, and the Life Course*. Weinheim, Germany: Deutscher Studien Verlag, 67–81.

Overington, C. 2012. If this baby had been born fifteen years ago, these photos would not exist. *Australian Women's Weekly* (October), 60–66.

Oxford University. 1971. *The Compact Edition of the Oxford English Dictionary*. Oxford: Oxford University Press.

Pahl, V.A. 1954. *Through the Iron Lung*. Thorold, Ontario, Canada: Dryden Sinclair

Pain, R. 2001a. Age, generation and the lifecourse. In Pain, R. (ed.) *Introducing Social Geographies*. London: Arnold.

Pain, R. 2001b. *Introducing Social Geographies*. London: Arnold.

Pain, R. 2005. Intergenerational relations and practice in the development of sustainable communities. Background paper, Office of the Deputy Prime Minister. https://
www.gov.uk/government/organisations/deputy-prime-ministers-office

Pain, R. 2006. Intergenerational relations and practice in the development of sustainable communities. Background paper for the Office of the Deputy Prime Minister,
London: ODPM.

Parkhurst, G., Galvin, K., Musselwhite, C., Phillips, J., Shergold, I. and Todres, L.
2012. A continuum for understanding the mobility of older people. Working paper.
Bristol: University of the West of England.

Passenger Focus. 2013. *Bus Passenger Views on Value for Money.* London.

Peel, C., Sawyer Baker, P., Roth, D., Brown, C., Brodner, E. and Allman, R. 2005.
Assessing mobility in older adults: the UAB study of aging life-space assessment.
Physical Therapy, 85(10), 1008–1119.

Peraldi, M. 2005. Les nouvelles routes algériennes. In Anteby, L., Sheffer, G. and
Berthomière, W. (eds) *Les diasporas: 2000 ans d'histoire.* Rennes, France: Presses
Universitaires de Rennes, 371–384.

Pfeil, U., Zaphiris, P. and Wilson, S. 2009. Older adults' perceptions and experiences
of online social support. *Interacting with Computers,* 21(3), 159–172.

Pfirsch, T. 2011. Une géographie de la famille en Europe du Sud. *Cybergeo: European
Journal of Geography,* 533. http://cybergeo.revues.org/23669

Pharmacopoeia. 2003. *Cradle to Grave.* London: The British Museum Wellcome Trust
Gallery.

Pheeraphuttharangkoon, S., Choudrie, J., Zamani, E. and Giaglis, G. 2014. Investigating
the adoption and use of smartphones in the UK: a silver-surfer's perspective. ECIS
2014 – 22nd European Conference on Information Systems. Tel-Aviv, Israel, 9–14 June.

Philippine Overseas Employment Agency. n.d. Deployed land based overseas workers by
destination. (no page) www.poea.gov.ph/stats/2011Stats.pdf (Accessed 11 April 2015).

Phillipson, C. 2004. Urbanization and ageing. *Ageing and Society,* 24, 963–972.

Pilcher, J. 1994. Mannheim's sociology of generations: an undervalued legacy. *British
Journal of Sociology,* 45. http://leicester.academia.edu/JanePilcher/Papers/244971/
Mannheim-S-Sociology-of-Generations–An-Undervalued-Legacy

Pink, S. 2009. *Doing Sensory Ethnography.* London: Sage.

Piper, A.M., Weibel, N. and Hollan, J.D. 2014. Designing audio-enhanced paper
photos for older adult emotional wellbeing in communication therapy. *International
Journal of Human-Computer Studies,* 72, 629–639.

Polkinghorne, D.E. 1998. *Narrative Knowing and the Human Sciences.* Albany: State
University of New York Press.

Ponzalesi, S. and Leurs, K. 2014. On digital crossings in Europe. *Crossing: Journal of
Migration and Culture,* 5(1), 3–22.

Pooley, C., Turnbull, J. and Adams, M. 2005. '… everywhere she went I had to tag
along beside her': family, life course, and everyday mobility in England since the
1940s. *History of the Family,* 10(2), 119–136. doi: 10.1016/j.hisfam.2004.11.001

Porio, E. 1999. Global householding, gender and Filipino migration: a preliminary
review. *Journal of Philippine Studies,* 55(2), 211–242.

Porter, G., Hampshire, K., Ababe, A., Munthall, A., Robson, E., Mashiri, M. and
Maponya, G. 2010. Where dogs, ghosts and lions roam: learning from mobile
ethnographies on the journey from school. *Children's Geographies,* 8(2): 91–105.

Porter, G. and Mawdsley, E. 2008. Mobility and development. *Geography Review,*
21(4), 16–18.

Power, A. 2012. Social inequality, disadvantaged neighbourhoods and transport deprivation: an assessment of the historical influence of housing policies. *Journal of Transport Geography*, 21, 39–48.

Prensky, M. 2001. Digital natives, digital immigrants, part 1. *On the Horizon*, 9(5), 1–6.

Prohaska, T., Anderson, L., Hooker, S., Hughes, S. and Belza, B. 2011. Mobility and aging: transference to transportation. *Journal of Aging Research*. doi: 10.4061/2011/392751

Pugliese, E. 2011. Il modello mediterraneo dell'immigrazione. In Miranda, A. and Signorelli, A. (eds) *Pensare e Ripensare le migrazioni*. Palermo: Sellerio.

Recchi, E. 2014. Pathways to European identity formation: a tale of two models. *Innovation: the European Journal of Social Science Research*, 27(2), 119–133.

Rhoades, R. 1978. Foreign labour and German industrial capitalism 1871–1978: the evolution of a migratory system. *American Ethnologist*, 5, 553–573.

Righi, V., Sayago, S. and Blat, J. 2012. Older people's use of social network sites while participating in local online communities from an ethnographical perspective. CIRN 2012 – Community Informatics Conference: 'Ideals Meet Reality'. Prato, Italy, November 7–9.

Riley, M.W. and RileyJr, J.W. 1994. Age integration and the lives of older people. *The Gerontologist*, 34, 110–115.

Riley, M.W. and RileyJr, J.W. 2000. Age-integration: conceptual and historical background. *The Gerontologist*, 40, 266–270.

Ritzer, G. 2004. *Sociological Theory*. 4th edn. New York: McGraw Hill.

Robertson, S. 2007. Visions of urban mobility: the Westway, London. *Cultural Geographies*, 14(1), 74–91.

Robles, A. 1986. Perception of parental nurturance, punitiveness and power by selected Filipino primary school children. *Philippine Journal of Psychology*, 19, 18–28.

Roger, B.W. 1984. *Studies in International Labour Migration*. London: Macmillan.

Rogers, E.M. 2003. *Diffusion of Innovations*. 5th edn. New York: Free Press.

Rosales, A., Righi, V., Sayago, S. and Blat, J. 2012. Ethnographic techniques with older people at intermediate stages of product development. NordiCHI'12 workshop: How to design touch interfaces for and with older adults: identification of challenges and opportunities. Copenhagen, Denmark, 14–17 October.

Rose, G. 2014. On the relation between 'visual research methods' and contemporary visual culture. *The Sociological Review*, 62(9), 24–46.

Rose, N. 1998. *Inventing Our Selves: Psychology, Power, and Personhood*. Cambridge Studies in the History of Psychology. Cambridge: Cambridge University Press.

Rosoli, G. 1976. *Un Secolo di Emigrazione Italiana 1876–1976*. Rome: Centre Studi Emigrazione.

Rothwell, J. and Massey, D. 2014. Geographic effects on intergenerational income mobility. *Economic Geography*, 91(1), 83–106.

Rudman, D.L. 2006. Shaping the active, autonomous and responsible modern retiree: An analysis of discursive technologies and their links with neo-liberal political rationality. *Ageing and Society*, 26(2), 181–201.

Rudulph, M. 1984. *Inside the Iron Lung*. Bourne End, Buckinghamshire: Kensal Press.

Sanfilippo, M. 2011. Il fenomeno migratorio italiano ; storia e storiografia. In Miranda, A. and Signorelli, A. (eds) *Pensare e Ripensare le migrazioni*. Palermo: Sellerio, 245–273.

Save the Children. 2011. *A Focus on Child Protection Within Social Protection Systems: Transforming Children's Lives*. Stockholm: Save the Children Sweden.

Sawatzky, R., Liu-Ambrose, T., Miller, W. and Marra, C. 2007. Physical activity as a mediator of the impact of chronic conditions on quality of life in older adults. *Health and Quality of Life Outcomes*, 5, 68.

Sayad, A. 1977. Les trois âges de l'émigration algérienne en France. *Actes de la Recherche en Sciences Sociales*, 15, 59–79.

Sayago, S. and Blat, J. 2010. Telling the story of older people e-mailing: an ethnographical study. *International Journal of Human-Computer Studies*, 68, 105–120.

Sayago, S., Forbes, P. and Blat, J. 2013. Older people becoming successful ICT learners over time: challenges and strategies through an ethnographical lens. *Educational Gerontology*, 39(7), 527–544.

Sayago, S., Sloan, D. and Blat, J. 2011. Everyday use of computer-mediated communication tools and its evolution over time: an ethnographical study with older people. *Interacting with Computers*, 23(5), 543–554.

Scabini, E. and Donati, P. (eds) 1988. La famiglia lunga del giovane adulto: verso nuovi compiti evolutivi. *Studi Interdisciplinari Sulla Famiglia*. Vol. 7. Milan: Università Cattolica.

Schäfer, M., Jaeger-Erben, M. and Bamberg, S. 2012. Life events as windows of opportunity for changing towards sustainable consumption patterns? *Journal of Consumer Policy*, 35(1), 65–84.

Scharf, T.Phillipson, C. and Smith, A. 2005. Social exclusion of older people in deprived urban communities of England. *European Journal of Ageing*, 2, 76–87.

Scheiner, J. 2007. Mobility biographies: elements of a biographical theory of travel demand. *Erdkunde*, 61(2), 161–173.

Scheiner, J. 2014. Residential self-selection in travel behavior: towards an integration into mobility biographies. *Journal of Transport and Land Use*, 7(3), 15–28.

Scheiner, J. and Holz-Rau, C. 2013. Changes in travel mode use after residential relocation. a contribution to mobility biographies. *Transportation*, 40(2), 431–458.

Schwanen, T. 2007. Gender differences in chauffeuring children among dual-earner families. *The Professional Geographer*, 59(4), 447–462.

Schwanen, T. 2015. Geographies of transport I: Reinventing a field? *Progress in Human Geography*, 40(1). Online first. doi:10.1177/0309132514565725

Schwanen, T., Hardill, I. and Lucas, S. 2012. Spatialities of ageing: the co-construction and co-evolution of old age and space. *Geoforum*, 43(6), 1291–1295.

Schwanen, T. and Páez, A. 2010. The mobility of older people: an introduction. *Journal of Transport Geography*, 18(5), 591–595.

Schwartz, M. 1996. *Morrie: In His Own Words*. New York: Walker and Company.

Scotto, G. 2015. From 'emigrants' to 'Italians': what is new in Italian migration to London? *Modern Italy*, 20(2), 153–155.

Sennett, R. 1970. *The Uses of Disorder: Personal Identity and City Life*. London: Allen & Unwin.

Sennett, R. 2006. The open city. Newspaper essay, Urban Age. London School of Economics. http://downloads.lsecities.net/0_downloads/Berlin_Richard_Sennett_2006-The_Open_City.pdf (Accessed 13 July 2015).

Sgritta, G.B. 2002. La transizione all'età adulta : la sindrome del ritardo. *Famiglie. Mutamenti e politiche sociali*. Bologna: Il Mulino.

Shakespeare, W. c.1600. *The Tragedy of Hamlet, Prince of Denmark*. Act III, Scene I. Hosted at the Shakespeare Quartos Archive as Hamlet, 1604. Copy 1. Folger Library, image 8. www.quartos.org/ (Accessed 13 December 2013).

Shaw, J. and Hesse, M. 2010. Transport, geography and the 'new' mobilities. *Transactions of the Institute of British Geographers*, 35(3), 305–312.

Shaw, B., Watson, B., Frauendienst, B., Redecker, A. and Jones, T. (with Mayer Hillman), 2013. *Children's Independent Mobility: A Comparative Study in England and Germany (1971 to 2010)*. London: Policy Studies Institute.

Sheller, M. 2008. Mobility freedom and public space. In Bergmann, S. and Sager, T. (eds) *The Ethics of Mobilities: Rethinking Place, Exclusion, Freedom and Environment.* Aldershot: Ashgate.

Sheller, M. and Urry, J. 2006. The new mobilities paradigm. *Environment and Planning A*, 38(2), 207–226.

Shergold, I. and Parkhurst, G. 2012. Transport-related social exclusion amongst older people in rural Southwest England and Wales. *Journal of Rural Studies*, 28(4), 412–421.

Silverman, D. 2011. *Interpreting Qualitative Data: A Guide to the Principles of Qualitative Research*. London: Sage.

Skeggs, B. 2004. *Class, Self, Culture*. London: Routledge.

Smith, R.C. 2002. Life course, generation, and social location as factors shaping second-generation transnational life. In Levitt, P. and Waters, M.C. (eds) *The Changing face of Home: The Transitional Lives of the Second Generation*. New York: Russell Sage Foundation.

Smithson, J., 2000. Using and analysing focus groups: limitations and possibilities. *International Journal of Social Research Methodology*, 3(2), 103–119.

Smythe, W.E. and Murray, M.J. 2000. Owning the story: ethical considerations in narrative research. *Ethics and Behavior*, 10(4), 311–336.

Social Exclusion Unit. 2003. *Making the Connections: Final Report on Transport and Social Exclusion*. London: SEU.

Social Exclusion Unit. 2006. *A Sure Start to Later Life*. London: SEU.

Somavia, J. 2010. Migrant workers are an asset to every country where they bring their labour. International Labour Organization. Available at www.ilo.org/public/eng lish//protection/migrant/

Sri, C. 2009. Gender, migration and social change: the return of Filipino women migrant workers. Unpublished doctoral dissertation. University of Sussex.

Stinner, W.F., De Albuquerque, K. and Bryce-Laporte, R.S. (eds) 1982. *Return Migration and Remittances: Developing a Caribbean Perspective.* Washington, DC: Smithsonian Institute, 45–73.

Stjernborg, V., Emilsson, U. and Stahl, A. 2014. Changes in outdoor mobility when becoming alone in the household in old age. *Journal of Transport and Health*, 1, 9–16.

Sumner, A., Haddad, L. and Gomez Climent, L. 2009. Rethinking intergenerational transmission(s): does a wellbeing lens help? The case of nutrition. *IDS Bulletin*, 40(1), 22–30 (Brighton: IDS).

Tan, A.L. 1989. Four meanings of fatherhood. *Philippine Journal of Psychology*, 22, 51–60.

Taylor, J., Bernard, M., White, C. and Lewis, J. 2007. *Understanding the Travel Aspirations, Needs and Behaviour of Young Adults*. London: Department for Transport.

Tee, K., Brush, A.J.B. and Inkpen, K.M. 2009. Exploring communication and sharing between extended families. *International Journal of Human-Computer Studies*, 67(2), 128–138.

Thang, L. 2001. *Generations In Touch: Linking the Old and Young in a Tokyo Neighbourhood*. Ithaca, NY: Cornell University Press.

The British Museum. 2014. Explore/Online tours: *Cradle to Grave* by Pharmacopoeia. www.britishmuseum.org/explore/online_tours/museum_and_exhibition/audio_descri ption_tour/cradle_to_grave_by_pharmacopoe.aspx (Accessed 10 February 2014).

Titheridge, H., Mackett, R.L., Christie, N., Oviedo Hernández, D. and Ye, R. 2014. *Transport and Poverty: A Review of the Evidence*. London: University College London.

Toscano, E.2011. Italian immigration in France: a never-ending phenomenon. *Altreitalie*, 43, 30–46.

Triandafyllidou, A. 2015. Reform, counter-reform and the politics of citizenship: local voting rights for third country nationals in Greece. *Journal of International Migration and Integration*. 16(1), 43–60. doi: 10.1007/s12134-014-0343-7. Published online on I-First, 28 April 2014. http://link.springer.com/article/10.1007/s12134-014-0343-7

Triandafyllidou, A. and Gropas, R. 2014. 'Voting with their feet' Highly skilled emigrants from Southern Europe, *American Behavioral Scientist*, 58(12), 1614–1633

Turner, M., Love, S. and Howell, M. 2008. Understanding emotions experienced when using a mobile phone in public: the social usability of mobile (cellular) telephones. *Telematics and Informatics*, 25, 201–215.

Uhlenberg, P. 1978. Changing configurations of the life course. In Hareven, T.K. (ed.) *Transitions: The Family and the Life Course in Historical Perspective*. New York: Academic Press, 65–87.

Uhlenberg, P. and Mueller, M. 2003. Family context and individual well-being. In Mortimer, J.T. and Shanahan, M.J. (eds) *Handbook of the Life Course*. New York: Plenum Publishing.

UK Government. 2013. *Super Connected Cities Programme*. Department for Culture, Media and Sport, UK Government. Available online at https://www.gov.uk/gov ernment/policies/transforming-uk-broadband/supporting-pages/investing-in-super-connected-cities (Accessed 10 March 2015).

United Nations. 2007. UN Committee on the Rights of the Child (CRC). UN Committee on the Rights of the Child: concluding observations, Kenya, 19 June 2007, CRC/C/KEN/CO/2. Available at: www.refworld.org/docid/4682102b2.html (Accessed 25 May 2015).

United Nations, Department of Economic and Social Affairs, Population Division. 2002. *World Population Ageing: 1950–2050*. New York: UNDESA.

United Nations, Department of Economic and Social Affairs, Population Division. 2013. *World Population Ageing 2013*. New York: UNDESA.

Uprichard, E. 2008. Children as being and becomings: children, childhood and temporality. *Children and Society*, 22(4), 303–313.

Urry, J. 2007. *Mobilities*. Malden MA and Cambridge: Polity Press.

Uvin, P., Jain, P.S. and Brown, L.D. 2000. Think large and act small: toward a new paradigm for NGO scaling up. *World Development*, 28(8), 1409–1419.

Valentine, G. 1996. Children should be seen and not heard: the production and transgression of adults' public space. *Urban Geography*, 17(3), 205–220.

Valentine, G. 2008. Living with difference: reflections on geographies of encounter. *Progress in Human Geography*, 32(3), 323–337.

van der Waerden, P., Timmermans, H. and Borges, A. 2003. The influence of key events and critical incidents on transport mode choice switching behaviour: A descriptive analysis. 10th International Conference on Travel Behaviour Research, Lucerne, Switzerland.

Van Maanen, J. 2011. *Tales of the Field: On Writing Ethnography.* Chicago and London: University of Chicago Press.

Vanderbeck, R. 2007. Intergenerational geographies: age relations, segregation and re-engagements. *Geography Compass*, 1(2), 200–221.

Vanoutrive, T. 2015. The modal split of cities: a workplace-based mixed modelling perspective. *Tijdschrift voor Economische en Sociale Geografie*, 106(5), 503–520.

Venkatesh, V. 2014. Technology acceptance model and the unified theory of acceptance and use of technology. In Cooper, C.L. (ed.) *Wiley Encyclopedia of Management.* 3rd edn. Chichester: Wiley.

Vergunst, J. 2010. Rhythms of walking: history and presence in a city street. *Space and Culture*, 14(4), 376–388.

Vetere, F., David, H., Gibbs, M. and Howard, S. 2009. The magic box and collage: responding to the challenge of distributed intergenerational play. *International Journal of Human-Computer Studies*, 67(2), 165–178.

Vincent, J. 2005. Emotional attachment to mobile phones: an extraordinary relationship. In Hamill, L., and Lasen, A. (eds) *Mobile World: Past Present and Future.* New York: Springer.

Vincent, J., Phillipson, C. and Downs, M. (eds) 2006. *The Futures of Old Age.* London: Sage.

Webber, S.C., Porter, M.M. and Menec, V.H. 2010. Mobility in older adults: a comprehensive framework. *Gerontologist*, 50(4), 443–450.

Wei, R. and Leung, L. 1999. Blurring public and private behaviors in public space: policy challenges in the use and improper use of the cell phone. *Telematics and Informatics*, 16, 11–26.

Weller, S. 2006. Situating (young) teenagers in geographies of children and youth. *Children's Geographies*, 4(1), 97–108.

Whyte, W.H. 1980. *The Social Life of Small Urban Spaces.* Washington, DC: Conservation Foundation.

Wiles, J. 2003. Daily geographies of caregivers: mobility, routine, scale. *Social Science and Medicine*, 57, 1307–1325.

Wilkinson, S. 1998. Focus group methodology: a review. *International Journal of Social Research Methodology*, 1(3), 181–203.

Williams, A. and Nussbaum, J. 2001. *Intergenerational Communication Across the Lifespan.* Mahwah, NJ: Lawrence Erlbaum Associates.

Wingens, M., De Valk, H., Windzio, M. and Aybek, C. 2011. *The Sociological Life Course Approach and Research on Migration and Integration: A Life-Course Perspective on Migration and Integration.* New York: Springer.

Wixey, S., Jones, P., Lucas, K. and Aldridge, M. 2005. Measuring accessibility as experienced by different socially disadvantaged groups. Working Paper 1, User Needs Literature Review. London: University of Westminster Transport Studies Group.

Wood, C. 1982. Equilibrium and historical-structural perspectives on migration. *International Migration Review*, 2, 298–318.

World Bank. 2006. *Global Economic Prospects: Overview and Global Outlook.* Washington, DC: World Bank.

Wright, E.O. 2010. *Envisioning Real Utopias.* London: Verso.

Wright, E.O. 2013. Transforming capitalism through real utopias. *American Sociological Review*, 78(1), 1–25.

Wu, Y. and Xu, H. 2015. Lifestyle mobility in China: context, perspective and prospect. Mobility Intersections conference. Lancaster University, Lancaster, July.

Wyly, E. 2011. Positively radical. *International Journal of Urban and Regional Research*, 35(5), 889–912.

Xie, B. 2008. Multimodal computer-mediated communication and social support among older Chinese internet users. *Journal of Computer-Mediated Communication*, 13(3), 728–750.

Yeoh, B. 2004. Cosmopolitanism and its exclusions in Singapore. *Urban Studies*, 41, 2431–2435.

Young, I.M. 1990. *Justice and the Politics of Difference*. Princeton, NJ: Princeton University Press.

Zaphiris, P. and Sarwar, R. 2006. Trends, similarities, and differences in the usage of teen and senior public online newsgroups. *ACM Transactions on Computer-Human Interaction*, 13(3), 403–422.

Zeitler, E., Buys, L., Aird, R. and Miller, E. 2012. Mobility and active ageing in suburban environments: findings from in-depth interviews and person-based GPS tracking. *Current Gerontology and Geriatrics Research*. http://dx.doi.org/10.1155/2012/257186

Zhang, J. 2014. Revisiting residential self-selection issues: a life-oriented approach. *Journal of Transport and Land Use*, 7(3), 29–45.

Ziegler, F. and Schwanen, T. 2011. I like to go out to be energised by different people: an exploratory analysis of mobility and wellbeing in later life. *Ageing and Society*, 31, 758–781.

Index